LABOR RELATIONS

for the Fire Service

LABOR RELATIONS for the Fire Service

PAUL J. ANTONELLIS JR.

PennWell®

Fire Engineering®

Disclaimer

The recommendations, advice, descriptions, and methods in this book are presented solely for educational purposes. The author and the publisher assume no liability whatsoever for any loss or damage that results from the use of any of the material in this book. Use of the material in this book is solely at the risk of the reader.

PennWell Corporation
1421 South Sheridan Road
Tulsa, Oklahoma 74112-6600 USA

800.752.9764
+1.918.831.9421
sales@pennwell.com
www.Fire EngineeringBooks.com
www.pennwellbooks.com
www.pennwell.com

Marketing Coordinator: Jane Green
National Account Manager: Cindy J. Huse

Director: Mary McGee
Managing Editor: Marla Patterson
Production Manager: Sheila Brock
Production Editor: Tony Quinn
Author photo by KS Michaud Photography

Library of Congress Cataloging-in-Publication Data

Antonellis, Paul J.
Labor relations for the Fire Service / Paul J. Antonellis, Jr.
p. cm.
Includes bibliographical references and index.
ISBN 978-1-59370-284-7
1. Fire fighters--Labor unions--United States. 2. Industrial relations--United States. 3. Collective bargaining--Fire fighters--United States. I. Title.
HD6515.F5A58 2012
331.88'11363370973--dc23
2012000436

Printed in the United States of America

2 3 4 5 16 15

Praise for *Labor Relations for the Fire Service*

This book will be of great use to labor relations practitioners in the public sector, fire service workers, and anyone who wants to learn more about what it really means to be employed in and lead fire services in the United States.

—Jason Russell, Ph.D., assistant professor of labor studies; coordinator, graduate program labor and policy studies, SUNY Empire State College

An insightful and historical perspective of labor relations in fire service which goes beyond what happens at the bargaining table. A must-read guide for both labor and management.

—Michael Sacco, Esq., Westfield, Massachusetts

Encapsulating a myriad of essential topics within labor relations and human resources, *Labor Relations for the Fire Service* is a progressive and realistic instructional tool. The book has broad application and use as either a text for courses at institutions of higher education or as reference guide in the firehouse.

—Elizabeth A. Shannon, fire prevention officer, Nantucket Fire Department, Local 2509

During my tenure as a firefighter and eventually as a chief officer, I found myself negotiating many labor contracts for several bargaining units over the years. I found this book to be an excellent presentation of the issues facing labor and management today. Paul hit the nail on the head with this book and it is a must read for all in the fire service.

—John K. Murphy, JD, PA-C, MS, EFO, Sammamish, Washington

Labor Relations for the Fire Service is a comprehensive text that should be required reading for every member of the fire service and administration, from the new recruit to the seasoned chief to the mayor, especially as the fire service evolves to confront the changes in the world. *Labor Relations for the Fire Service* is a text that should be found on every chief officer's bookshelf.

—Joseph C. Stringfellow, Esq., chief, Special Operations and Emergency Response UMDNJ, Camden, New Jersey

To my brother, Joseph K. Davini

Contents

Preface

Intent

Labor Relations for the Fire Service is intended as a textbook for college-level classes in the fire science program and meets the Fire and Emergency Service Higher Education (FESHE) requirement established by the U.S. Fire Administration for the course "Personnel Management for the Fire and Emergency Services." However, this textbook also serves as a useful tool for different levels within the fire service—from the line firefighter, union member, or volunteer to the company officer or fire service administrator. Many will find this text helpful for exploring and understanding the dynamics of labor relations as it pertains to the fire service.

To begin to understand labor relations, we must first identify the meaning of labor relations. For the purpose of this book, labor relations comprises the process that involves how the employer interacts with the employees and employee groups within the organization. This means that labor relations is about the relationship between the employee and the employer. When most people think of labor relations they think of unions, employers, bargaining, collective bargaining agreements, strikes, working conditions, wages, safety, seniority, working conditions, and grievance administration. However, labor relations is not just a union issue; as you will see, many of the same issues relate to nonunion work environments. Labor relations involves a great deal of human resource management skills and theories, and these will be explored in the following pages. We will explore the aspects of labor relations that are intertwined with human resource management in today's fire service organization.

A Desire to Write

During my more than 20 years in the fire service, I have had the opportunity to travel nationally and internationally, speaking on a number of fire service topics. During my travels I have found that, in general, the fire service is lacking a solid labor relations textbook that deals specifically with fire service issues.

When I taught at Anna Maria College, Southern New Hampshire University, and Empire State College, I lacked a labor relations textbook that fit the needs of the firefighters in today's fire service. Although there are some very well-written books pertaining to labor relations (focused on private sector and public sector) and public sector labor relations, what I quickly found was that many of these textbooks have not dealt specifically with today's fire service issues and that several of the chapters did not apply to the fire service; as a result, material was not being used, and students were getting lost in the material that did not apply to the fire service. Seeing the need for such a textbook, I set out on a three-year journey to research and write such a book. I am happy to say this book is the end product of many hours of research, conversations, and experience—and of the true enjoyment that I receive from teaching labor relations.

Labor Relations for the Fire Service introduces students to the labor relations process, including labor history, laws, negotiations, contract administration, discipline, personnel management, and the collective bargaining process and practices. My 20-plus years in the fire service—holding union positions; serving as fire chief for the Town of Salisbury (Mass.), volunteer firefighter with the Meadowood County (N.H.) Area Fire Department, and call firefighter with Winchendon (Mass.) Fire Department; holding career status with the Yarmouth (Mass.) Fire Department; serving as the Municipal Training Support Coordinator at the Massachusetts Firefighting Academy; and teaching more than 60 undergraduate- and graduate-level labor relations classes—provides the basis for this text. I have spent the better part of two years researching, writing, and editing this book. This book is far more than an "opinion" book; rather, references are cited to support my positions. It is my hope that this book will help to fill the literature gap in the area of labor relations in the fire service. Every attempt was made to balance labor and management issues throughout the book while attempting to touch on the challenges that call and volunteer fire departments may face with labor relation issues. A number of experts from different areas of the fire service—fire service administrators, union officials, labor attorneys, mid-level fire service managers, and academic scholars—have been asked to review this book and offer short quotations on the topic's importance.

I am always interested in what people think of the material contained within this book and look forward to receiving feedback from the readers. If you have any suggestions or comments on how to make this book better, please feel free to e-mail me at Paul.Antonellis@comcast.net. I am also available for speaking engagements, nationally and internationally, on labor relations issue or other fire service topics.

Acknowledgments

First I would like to thank PennWell Publishing for giving me this wonderful opportunity and for having faith in the project. Thank you to all of the staff and production personnel at PennWell who have worked alongside me to ensure the best quality of this book. It has been a long, strange, and bumpy road, but we did it! A special thanks goes to my friend, labor attorney, and mentor Michael Sacco. The greatest appreciation and thanks goes out to my loving family, who tolerated the many hours of research, writing, and editing that took away time from them. Above all, thanks goes to my loving wife, Susan, who is my foundation, inspiration, and reality checker. Susan has read and reread each chapter over the past two years, and without her consultation, support, and encouragement, I could not have taken this project to the finish line. Last of all, to my children, Joseph, Benjamin, and Rebecca, who popped in during those periods of writer's block, not only helping to break that block but also reminding me of what is truly important in life: family. So, to my family, I love you!

FESHE Outcomes Correlations

Course requirements for Personnel Management for the Fire and Emergency Services course developed as part of the Fire and Emergency Services Higher Education (FESHE) model curriculum.

FESHE Outcome	Chapter									
	1	2	3	4	5	6	7	8	9	10
	Page no(s).									
Identify and explain contemporary personnel management issues.	1–4	20–21 24–28	33–48	73–99	104–107 108–109 119–122	129–138 149–155	151–179	184–197 198–214	220–246	251–255 256–258 263–275
Explain potential personnel management issues.		8–19 24–28	43–48	73–99	104 107 113–116 119–122	129–138 140 153–155	151–179	184–214	220–246	251–256 256–258 260–263
Classify the collective rules, procedures, laws, and policies that relate to personnel management issues.		20–23	34–43 46–51 54 56 58–62	83–86 87–91	112–113 119	128 138–143 146–148 150–155	161 163–179	191–193 205–212	227–235 237–238 239–246	253 257–258 260 266–270
Analyze simple/ complex personnel management issues from recruitment to retirement.		20–21	39–43 52–56 62–68	73–75 88–91	113–119	131–137 138–145	151–179	183–214	220–221 222–223 224–226	256–258 259–270 270–272
Formulate recommendations and solutions to personnel management issues.		24–28	52–66	82 86	118 104–110	128 139 145	172	196–197 212	224 226 232 246	260
Explore organizational development and leadership styles and how they relate to personnel relationships.			53–57 58–62 66–67	92–99	104–106 119–123		161–165 171–179	183–214 216–217	233–246	251–256 258–260 263–272 273–276

SETTING THE STAGE FOR LABOR RELATIONS

1

Learning Objectives

After reading this chapter, you should be able to:

- Discuss the mix of personnel management and human resource management as it applies to labor relations in the fire service.
- Explain the chapter transitions and the layout of the book.
- Describe how this textbook is tailored to meet the needs of the labor relations student.

Introduction

This basic textbook will help potential fire service administrators, supervisors, union members, students, and all types of fire service personnel better understand the impact that labor relations has on the organization. The field of labor relations is very dynamic, so staying current with changing employment laws and court rulings is critical to the field of labor relations. With the correct mix of labor and management, the field can be very rewarding. When the balance is off, however, labor relations can become an unpleasant experience.

Because the fire service cannot function without the people who provide the service to the community directly and indirectly, labor relations must be part of the fire department. Labor relations should be part of officer development programs and should be included in degree plans at colleges. The fire department will be judged by the personnel within the organization and the quality of service provided. Personnel management goes hand in hand with labor relations. Human resource management goes hand in hand with labor relations. This text will provide you with a blend of human resource management and personnel management. Because the fire service today is made up of baby boomers, Generation Xers, and Millennials, fire service administrators need to understand the very different needs and expectations of each generation. The challenge that personnel management, human resource management, and labor relations will face is how best to meet these needs and expectations while finding creative ways to motivate, satisfy, reward, and recognize each generation.

Chapter Transition

To understand and appreciate the labor relations field fully, one must first recognize how we arrived at our current position. The field of labor relation started in the private sector years before the public sector was organized. In the 1790s the first craft union was formed (private sector), and it was not until the 1950s that the public sector unionism grew in strength and was legitimized. As we review the historical events that have helped shape today's labor relations field, you will find that many of the current labor laws have deep roots. It is these historical events that shaped labor relations in the private and public sectors. Because labor relations started well before the public sector was organized, we will begin by exploring the historical events of the labor relations field. Labor-management relations have evolved beyond the early days of violent confrontations, and today labor-management relations seek to build and foster cooperation and trust.

Chapter 3 (Labor-Management Relations in the Public Sector) explores what distinguishes labor relations between the public and the private sectors. Labor-management relations in the public sector are a very distinct animal compared to labor-management relations in the private sector. Public sector labor relations is an outgrowth of the private sector labor relations field. In the public sector, the labor relations laws, ordinances, regulations, and executive orders are crafted by individual states, rather than by a single federal law, such as the National Labor Relations Act or the Federal Labor Relations Act. A significant difference exists between public sector and private sector labor relations, even while there are recognizable similarities. Regardless of your rank or title, labor relations will have an impact on you and your department. Labor relations is not just about labor unions; it is about dealing with all types and forms of personnel issues. This means that labor relations must be part of career, call, volunteer, and private fire brigades. Regardless of whether you have people working for you or you are working for a person, labor relations will affect you, and you should have a clear understanding of your rights and privileges.

Chapter 4 (Labor Relations and Collective Bargaining) explores the collective bargaining process in the public sector. Individual state laws define the limits of collective bargaining for firefighters. State statutes or local ordinances may allow collective bargaining, in which case the law/ordinance will govern the relationship between the labor union and the employer. In other cases, state law does not allow collective bargaining with labor unions. Chapter 4 also breaks down the contract components commonly found in collective bargaining agreements in the fire service. Further, the scope of bargaining in the public sector is explored, dealing specifically with bargaining power, structure, and decentralized bargaining. For the bargaining process to be initiated, the legal rights of both the employer and employee must be understood.

A small percentage of firefighters will be involved in the actual negotiation process of a collective bargaining process. Many fire service administrators will be directly or indirectly involved in the negotiation process. Chapter 4 examines the process of how labor and management will prepare to come together to negotiate a collective bargaining agreement. Today, public sector bargaining laws are under significant review, and in many cases public sector bargaining laws are being changed in favor of the employer, weakening the bargaining power of public sector unions.

Chapter 5 (Negotiation Process and Impasse) examines the negotiation process commonly used in the fire service. Any time negotiations take place between two parties, there exists the risk of a bargaining impasse. This section will explore how impasses are commonly handled, such as mediation and arbitration.

Once the bargaining process has concluded, labor and management must each work to administer the collective bargaining agreement on a daily basis. Chapter 6 (Contract Administration) reviews the duties and responsibilities for labor and management. In personnel management, there should always be an employee dispute-resolution process to allow employees to resolve any personnel issues with the employer. It should not matter if you are career, volunteer, or on call, a dispute-resolution process needs to be included in your personnel management system. Chapter 6 examines the sources of employee grievances and the arbitration process that are commonly used to settle differences.

When dealing with employees, disciplinary actions must be part of the personnel management system. Regardless of the size of the fire department, there remains a chance that at some point disciplinary actions may need to be taken. Chapter 7 (Disciplinary Action) will review the concept of due process and the dynamics of the discipline process. The individual firefighter, as well as the fire service administrator, need to understand that employees have rights that must be protected and respected during a disciplinary action. Today, more than ever before, the fire service administrator must conduct a threat assessment of the department before taking any disciplinary action. The fire service is no longer immune from workplace aggression and violence. We will review the process used to assess the organization and steps to prepare for a suspected act of violence or aggression.

Regardless of the size or the makeup of the fire department, human resource management is a critical element for properly managing people. Chapter 8 (Human Resource Management) discusses the importance of human resource management and the central role it plays in the labor relations process. Human resource management is one of the most misunderstood functions in the fire service and is often overlooked with negative results. An in-depth review of the six key human resource functions is provided: planning, organizing, staffing, controlling, leading, and direction. Chapter 8 examines the types of personalities found in the fire service. Having a basic understanding of various personality traits aids the company officer and fire service administrator in managing firefighters. A connection can be drawn between the personalities and the motivations of firefighters. Properly managing human resources yields the greatest good for the organization. When an organization fails to manage its human resources, it essentially fails to achieve the greatest good for the organization. In addition, failing to manage human resources may put the organization in violation of employment laws and/or employees' rights and privileges. No matter the size or type of fire department, the recruitment and selection process presents challenges, and not selecting the right person for the position can have a costly impact, affect staff morale, and waste resources. Often fire departments overlook employee retention and career development programs. Once a firefighter is hired, the fire department should have a career development path for each position within the department. Chapter 8 discusses recommendations and solutions pertaining to the recruitment, selection, and retention process in the fire service.

Chapter 8 also discusses the *human resource triad*—that is, how fire service administration, human resource management, and the employees should all have equal input into the labor relations process. Each of the three components of the triad has some input into the process, utilizing the expertise of employees, fire administration, and human resource personnel.

In today's fire service, firefighters, officers, and fire service administrators are all affected by workplace stress and health concerns. The responsibility for ensuring firefighter health and wellness rests with the administration, supervisor, coworkers, and the individual firefighter. Many labor organizations have crafted various support programs for firefighters who have been traumatized on the job. Chapter 9 (Stress and Psychological Support) delves deep into the various support systems that are commonly available to fire service personnel and their families, such as employee assistance, wellness, and fitness-for-duty programs. The focus of this chapter is preservation of firefighters' careers. In some cases, firefighters may never be able to return to work as a direct result of injury on the job. This chapter discusses recommendations for assistance for members who are faced with forced retirement. The issue of a line-of-duty death/injury is discussed not only in terms of the impact it has on the individual employee and the organization but also, and most importantly, in terms of the effect that it has on the family. This is one issue that fire service management books have tended to overlook or downplay. Nevertheless, a line-of-duty death/injury can have a profound impact on the labor relations process.

Finally, chapter 10 examines the future of labor relations. To speculate on what tomorrow's labor relations will look like, one first requires a clear understanding of the past and what has brought us to this point as well as a thorough understanding of how the labor relations process currently works. Chapters 2 through 9 set the stage for how labor relations should begin to move forward.

Final Thoughts

Some of the chapters contain student activities that can aid in discussion of critical issues within each chapter. The student activities can be treated as short written assignments or as discussion points for the class. It is important that these activities be incorporated into the learning process to allow the student to better define the challenges, recommendations, and solutions to personnel management in the fire service. Each chapter closes with a section called "Exploring Issues on the Web," which includes a list of helpful Web sites to further explore that chapter's topic. Extensive references have been supplied to emphasize that the ideas presented in this textbook represent not just one person's opinion; rather, the ideas, recommendations, and solutions presented herein have the support of data and research. This book is intended to fill the needed gap in the literature and make up for the lack of labor relations information pertaining specifically to the fire service. Training and education programs need to be crafted for fire service administrators to work effectively with labor unions, call firefighter organizations, and volunteer fire companies. Leaders of these organizations also have a responsibility to learn how best to interact with administration. Learning as you go can be a time-consuming and costly endeavor.

HISTORICAL EVENTS IN LABOR RELATIONS AND THE FIRE SERVICE

2

Learning Objectives

After reading this chapter, you should be able to:

- Identify significant historical events that helped shape the labor movement.
- List labor organizations that have played a key role in developing the labor movement, together with their contributions.
- Describe the importance of labor laws.
- Explain the impact that the labor movement has had on the fire service.

The significant labor events of the past form the foundation of this book and will enhance our understanding of labor relations today. Historical events have had a profound impact on labor-management relations. The following sections are not intended to make you an expert on the history of the fire service; rather, they are intended to increase your awareness of the events that have shaped labor-management relations today.

Often, firefighters have a limited knowledge of labor relations, but labor relations plays a significant role in the fire service. Some union firefighters lack a true understanding of the labor movement from a historical standpoint. Many firefighters (full-time, call, and volunteer) complain about poor management, lack of pay raises, and safety issues, and the list goes on. What firefighters often do not realize is that their current labor issues are minor compared to what many people faced in the early days of the labor movement. In some cases, labor leaders gave their lives in the pursuit of better working conditions, many of which are still enjoyed today by workers across the United States. These better working conditions have benefited union and nonunion workers. The study of labor relations is not just about organized labor; rather, it is about how labor and management can work together for the benefit of the organization.

Some say labor relations is shaped by the following five areas: social issues, how employees/employers view the work environment, various classes, economics, and finally, power. Many of the issues covered in this book can be linked back to these five areas. As you are reading, see if you can link the topic being discussed back to one of these concerns.

This chapter lays the foundation for rest of the book. The historical events that have had an impact on labor relations have many connections to today's issues, approaches used, and laws. In contrast to many labor history texts, the following sections are organized by the importance of historical events to labor relations, rather than chronologically.

Fire Brigade

Firefighting became a formalized organization in the United States in the late 1700s. Many associate Benjamin Franklin with developing the country's first volunteer fire brigade. The first volunteer fire department began in Philadelphia in 1736 and was called the Union Fire Company. In 1737, the general assembly created the volunteer Fire Department of the City of New York, and the volunteer department served until after the Civil War. Around 1865, the Metropolitan Fire Department was established with paid firefighters to protect the city. However, in 1678, Boston became the first city to purchase a motorized fire apparatus, house it in a firehouse, and pay a person to man the rig.[1] The city of Chicago hired paid firefighters in 1858. In 1859, the city of Baltimore hired paid firefighters to staff one truck and four engine companies. In December of 1866, the San Francisco fire department moved from an all-volunteer department to hiring its first paid firefighters; the pay rate for the chief engineer was $4,000 a year, and the three assistant chiefs were each paid $50 a month. It was not until 1900 that the department transitioned to an all-paid department and the call firefighter's force had been abolished.[2] In May 1889, the city of San Diego, California, drafted a city constitution for a new municipal government. The charter proposed the formation of a paid fire department.[3] The following prominent names in American history all served at one time as volunteer firefighters: George Washington, Thomas Jefferson, Benjamin Franklin, Samuel Adams, John Hancock, Paul Revere, Alexander Hamilton, and Benedict Arnold. The volunteer fire brigade has played a significant role in the American fire service historically, and it still does today.

Historical Labor Movement

In 1806, the United States presided over its first significant labor ruling, Commonwealth v. Pullis, also known as the Philadelphia Cordwainers case. Under English common law, two or more employees could not organize with the intent to pressure an employer to increase their wages or improve their working conditions. Philadelphia shoemakers attempting to do just that were indicted, convicted, and assessed $8. At hand was a group of employees attempting to regulate the price of labor of others, as well as their own. Shop owners feared the worst: if an uprising of employees could control the cost of labor, then they could be forced to close their doors.[4] During this period unions did not stand a chance. The unions not only had to fight management but also the federal government. The Cordwainers case resulted in a landmark decision that set the precedent for many years to follow. The shoemakers struck for higher wages because management refused to listen to their demands. The union leaders were indicted for the crime of conspiracy. The jury in the Cordwainers case consisted of businessmen, and the judge sitting on the case was a Federalist, neither friendly toward labor.

In the 1800s, many people left family farms to work for other people, especially in mills. This was a new process and was referred to as America's Industrial Revolution. The Industrial Revolution was when people made machines to produce a product/service that customers wanted. This was the birth of mills, plants, and factories in the United States—something never witnessed before.

With change would come resistance. Instead of many individualized craftspeople performing a trade in their homes and shops, they would now be in a centralized location and comprise a working organization in mills, plants, and factories. Harry R. Carter and Erwin Rausch state that "the Industrial Revolution increased specialization and gradually resulted in the development of machines that replaced people. Sometimes, several of the more sophisticated machinery could be tended by one little girl and often were. Child labor laws had yet to be enacted."[5] Labor would begin its long fight for dignity, justice, and economic security in the work environment. This is a fight that many unions and employees are still waging today.

The introduction of machines also had an impact on the fire service. Old leather buckets would be replaced by fire pumps. In 1914, the Bakersfield Fire Department was the first department in California to become completely motorized.[6] In 1922, the San Francisco Fire Department replaced horse-drawn steam engines with motorized fire apparatus. The introduction of motorized fire equipment reduced the number of people needed to flow water onto a fire and increased the amount of water flowing, changing the fire service's approach to fire extinguishment (fig. 2–1).

On the labor front, in 1818, the fire service added its first known female volunteer firefighter, an African American slave named Molly Williams. It was not until the mid-1850s that the trend became for cities to pay people to staff fire stations. In New England, the smokestacks of new mills could be seen in many of the cities and manufacturing towns. Industrialization spread very quickly across America, transforming the face of labor.

A main issue labor organizations focused on was child labor. In 1836, Massachusetts passed the first labor law that covered children. In part, the law stated that children under the age of 15 could not work in incorporated factories, unless the child had attended school for three months prior. Around the mid-1830s, the railroads in the country were experiencing tremendous growth, and thousands of men were employed by the railroad industry. During this time, the government and the courts did not take an active role in the growth of the industry. As a result of this hands-off approach, economic wealth and power were bestowed on the industrialists. The focus was on the nationalizing of business, turning a blind eye to the labor force in the United States.

The fire service would be no different. As cities began to hire paid firefighters, it could be argued that the government exploited firefighter labor with long hours, low pay, and unsafe working conditions. The early paid firefighters worked an 84-hour workweek; firefighters truly lived and worked at the fire station, with short breaks to visit their families. It has been reported that the pay was as low as 29 cents per hour in some cities. Because the industry of firefighting lacked any real history, equipment and safety were not always concerns. The industrialist viewed labor as a commodity—a commodity they wanted to obtain as cheaply as possible.

On July 4th, 1864, Congress passed the Contract Labor Law. The act was developed to encourage immigration, and created the U.S. Immigration Office. The act provided a means for workers from Great Britain, France, Switzerland, Norway, and Sweden to work in the United States in turn for a lien on their wages. In simple terms, this act provided the immigrants a way to work in the United States in turn for prepaid passage from their homelands (i.e., contract laborers). The Contract Labor Law was repealed in 1868, putting an end to the government's attempt to introduce immigrant workers into the U.S. labor force. Even after the law was repealed, some companies continued the practice of recruiting immigrants, which could be seen in the construction of the First Transcontinental Railroad. The railway industry brought Chinese labor to the United States to work on the construction of railways.[7]

Fig. 2–1. Orange (CA) Fire Department truck No. 1, 1912. (Photo courtesy Orange County Archives.)

In the early 1900s, some paid firefighters decided that they needed the protection of a union to work on their behalf for better working conditions, pay, and safety. Washington, D.C., had the first fire department to join the American Federation of Labor (AFL). In 1903, Pittsburgh firefighters joined the AFL (Local 11431) and would later become Local 1 of the International Association of Fire Fighters (IAFF). The IAFF was officially formed in 1918 with the help of Samuel Gompers, the president of the AFL (fig. 2–2). Gompers saw the growing need for a specialized union to organize the growing number of firefighters in the United States.

In the 1950s, Richmond, Virginia, hired its first African American firefighter. The department continued to hire black firefighters, yet they were not allowed in three of the city's fire stations (6, 13, and 19). It was not until 1974 that the fire service hired a woman as a paid firefighter, in Arlington County, Virginia. Judith Livers worked her way up the ladder and eventually retired at the rank of battalion chief in 1999. According to Deputy Chief John Gelinas of the Cambridge, Massachusetts, Fire Department, Patrick H. Raymond was the first black fire chief in the United States. Chief Raymond was appointed to the department in the early 1850s and was appointed chief engineer (chief of department) on January 5, 1871.[8]

Fig. 2–2. Samuel Gompers, president of the AFL between 1886 and 1924

Around 1869, Uriah Smith Stevens established a fraternal order of tailors in Philadelphia. By the mid-1880s, the Noble and Holy Order of the Knights of Labor had a membership in excess of 700,000 members nationally. Membership was no longer just tailors, but included various skilled and unskilled occupations. The Knights of Labor accepted women and African Americans after 1883; it was not just a white male labor organization. At the time, the Knights of Labor was one of the first labor organizations large enough to take on industry and represented the issues of the average working person on a national front. The Knights of Labor fought for the eight-hour day, advocated to end child labor, and sought equal pay for equal work. The group was formed around the idea of unity and solidarity of working people. In an effort to maintain the safety of its members, membership in the Knights of Labor was kept secret. Members feared employer reprisal, an issue that unions still fight today.

Mary Harris "Mother" Jones (1837–1930) was a dressmaker in Chicago who lost her home during the Great Fire of 1871. She began attending the Knights of Labor meetings and quickly became an advocate for the working person. Mother Jones worked to organize the United Mine Workers and traveled from city to city supporting strikes; as she gathered support for the Knights of Labor, Mother Jones actively recruited women. During her travels, she faced violence and even jail time in her pursuit of labor ideals.[9]

The Federation of Organized Trades and Labor Unions (FOTLU) was formed in 1881 by members of the trades, including ironworkers, steelworkers, carpenters, printers, and glassworkers. The FOTLU and the Knights of Labor began with a working relationship that would later decline. In a reorganization, the FOTLU became the AFL in 1886. The AFL's formation resulted in more workers leaving the Knights of Labor. By 1890, the Knights of Labor membership fell to 100,000. Samuel Gompers served as the first president of the AFL. Gompers called on all unions to go out on strike wherever the eight-hour day was refused. It is estimated that 340,000 workers nationally took the fight to the strike lines over the refusal of an employer to move to the eight-hour day.

President Gompers was responsible for the AFL's continued growth. Meanwhile, the Socialist Labor Party and the Knights of Labor witnessed decreasing numbers. By around 1893, the Knights of Labor had more or less come to an end, a result of people taking a stand against labor agitators. The Knights of Labor gave way to the AFL and ceased operation by 1917.

In January 1875, a group of northeastern Pennsylvania miners organized as the Workingmen's Benevolent Association (Miners and Laborers) (WBA) voted to strike against the mine owners. The strike lasted six months, until the employees could no longer provide for themselves and their families. It was out of necessity that some of the employees returned to work. A group called the Molly Maguires pressured the owners of the mine with intimidation, physical assaults, arson, sabotage, and murder. The Molly Maguires felt that they had no choice but to act out because of unfair treatment by their employers, agitation by law enforcement and politicians, and ethnic discrimination.

The Philadelphia and Reading Railroad, which controlled most of the mines, hired Pinkerton Detective James McParland to infiltrate the Molly Maguires. In the end, the testimony of James McParland and others turned state's witnesses would result in conviction of 24 Irish miners; 10 were hanged for murder, and the others were sentenced to jail.[10] Kevin Kenny, in *Making Sense of the Molly Maguires*, explains that "the Molly Maguires always existed on two related levels: as a sporadic pattern of violence engaged in by a specific type of Irishman, and as a ubiquitous concept in a system of ideological representation."[11]

As far as this event goes, Franklin B. Gowen, president of the Philadelphia and Reading Railroad, won the long strike—but at what cost? Gowen was able to hire a man to get inside the internal workings of the union and to gather much needed evidence for the later prosecution of members of the Molly Maguires. Additionally, the Philadelphia and Reading Railroad was able to tap state powers; that is, the railroad paid the government for police power to aid it in overthrowing the union. Gowen established the Coal and Iron Police. At this point, the railroad had its own law enforcement agency, and state law enforcement agencies were subject to its influence. Any time the miners would hold a public protest, the Coal and Iron Police would respond with backup from the local law enforcement agencies.

During the trials (1862–1875), Irish Catholics were not allowed to sit on the jury. The Catholic Church played a significant role in the lives of Irish miners, so not allowing any Irish Catholics on the jury was deliberately discriminatory. The prosecution was led not by the state, but rather by attorneys for the Philadelphia and Reading Railroad. The investigation and arrest of the members of the Molly Maguires was conducted by the Philadelphia and Reading Railroad, utilizing the private investigation firm Pinkerton and the Coal and Iron Police. This was the first time that the state released its powers to a private firm—and with tragic results. The Molly Maguire incident was the first major labor event that resulted in death and injuries due to the labor movement. Many people still associate the Molly Maguires with union activities, but disagree about whether the Molly Maguires was in fact a union.

The WBA had been formed in 1868, only to become another casualty of the long strike, collapsing in 1875. The events surrounding the long strike were the public's first exposure to worker organizations/unions being connected with violence and criminal activities.

Both the WBA and the Molly Maguires had believed that their work was in the best interest of labor, and each group provided a social organization for people with like-minded views. Today, union members are still fighting the same issues: unfair treatment by employers, political pressure/power, and discrimination.

Haymarket Riot

During the 1870s and the 1880s, Chicago was the home to much labor unrest, union activists, and anarchists. Throughout the early part of 1886, the labor unions started the fight for an eight-hour day. During this time, employees would typically work 10–12 hour days, six days per week, for very little pay. The poor working conditions, low wages, and long workdays eventually brought things to a head. The issues were not new; employees had been attempting to change these conditions since the end of the Civil War.

On May 1, 1886, a nationwide strike for a shorter workday began. Two days later, at the McCormick reaper plant, a shooting took place during a riot, when police clashed with strikers. The shooting left one person dead.

On May 4, 1886, the union activists called for a protest meeting at Haymarket Square to protest the police actions at the McCormick plant days earlier. Mayor Carter Henry Harrison had issued a parade permit for the gathering. It was reported that about 2,500 people showed up at Haymarket Square that night. Chicago Police Inspector John Bonfield called in an additional 700 police officers to deal with the expected crowds and possible riots. The meeting and the attendees seemed orderly with no cause for concern. However, under pressure from business leaders, the police ordered the crowd to disperse at around 10:30 p.m. It was reported that an unknown person tossed a homemade pipe bomb into the crowd of more than 200 police officers. One officer was killed instantly (fig. 2–3), and six police officers were mortally wounded. Fellow officers started to shoot into the crowd and the fleeing spectators. In the end, 67 police officers were injured, and four workers were killed, and 50 or more were injured.[12]

Fig. 2–3. Haymarket Riot in 1886—a bomb exploding among the police in the street. Officer Mathias J. Degan was killed by the blast.

In the days and weeks after this event, the Chicago police harassed and terrorized the working class. It has been reported that hundreds of people were arrested by the police in connection with the bombing and that many were beaten. As Troy Taylor wrote, "eight conspirators were brought to trial for the riot and seven of them received the death sentence, while the eighth was given 15 years in prison."[13]

Because the riot led the public to perceive the union as an organization that would react with violence and death unless they get their way, the fight for the shorter day was lost. Behind the scenes, business leaders pressured law enforcement to break the unions. This represents another example of the loss of life families endured for the pursuit of better working conditions.

During the late 1890s, the United States saw the Homestead and Pullman strikes. At the time, these two labor actions had been the most significant the country had seen in some time. The depression of the 1890s made the industrial strife situation even more fragile. The Homestead strike and lockout took place in July 1892 at Andrew Carnegie's steelworks in Homestead, Pennsylvania. The company retained the protective services of the Pinkerton Agency (more than 300 armed men) as strikebreakers. During the strike a gun fight erupted, and in the end several people were left dead and many more injured. The U.S. government dispatched the state militia to assist the company in breaking the strike. Consequently, the Amalgamated Association of Iron and Steel Workers suffered a significant blow, as did the labor movement in the United States. By 1903, most steel companies in the United States had beaten down the unions, and, lacking any power, the unions no longer spoke for the workers.

The Pullman strike began because the Pullman Palace Car Company decided to cut employee wages by more than 25%. The American Railway Union (ARU) voted to go on strike as a result of the wage cuts and other factors. On May 11, 1894, the members of the ARU stopped working. The strike and the later boycott, in June 1894, together resulted in the interruption and, finally, the stoppage of rail service west of Chicago. More than 125,000 employees quit their jobs rather than touch a Pullman car. The U.S. government, under the direction of President Grover Cleveland, sent U.S. Marshals and the U.S. Army (2,000 troops) in to restore order and support the Pullman Palace Car Company. In the end, 13 people were killed and more than 55 people were injured. The Pullman strike served as yet another considerable blow to the labor movement in the United States.

After the Industrial Revolution, trade employment was growing in the United States, and the working class desperately needed representation. The increase in trade societies would clearly define the dividing line between labor and management in the U.S. The Wagner Act (National Labor Relations Act) was implemented in 1935 to guarantee that:

> Employees shall have the right to self-organize, to form, join, or assist labor organizations, to bargain collectively through representatives of their own choosing, and to engage in other concerted activities for the purpose of collective bargaining or other mutual aid and protection, and shall also have a right to refrain from any or all such activities.[14]

The greatest change to labor relations in this country was seen during the Depression years, specifically with the creation of the Wagner Act. Employees wanted representation in the workplace. The thinking was that the Wagner Act would protect the employees seeking union representation. The Wagner Act will be discussed further in later chapters.

The United States began the process of enacting labor laws with the passage of the Erdman Act of 1898 and the Clayton Act of 1914. Congress enacted the two laws in an effort to balance bargaining power between labor and owners. The Erdman Act made it a criminal offense for owners of companies to discriminate against or dismiss employees because of their membership in a labor union. The Erdman Act is also known for mediation and arbitration proceedings, the establishment of the voluntary federal arbitration board (it has ties to the Arbitration Act of 1888), and the banning of yellow-dog contracts.[15] The Erdman Act also provides protection for unions to self-organize and collective bargaining rights. The Erdman Act covered the operating brotherhoods, telegraphers, and switchmen for the railroad. From 1914 forward, labor would begin to see an era that supported the labor movement.

The Erdman Act has direct ties to the Pullman strike. After the strike, management for the Pullman Palace Car Company used a special blackball list that prevented employees from returning to work at the company. Some of the actions employed by managers included not providing employees (strikers) with letters stating their past work experience in the railroad industry. Without such a letter, an employee was almost guaranteed not to get a job elsewhere in the railway industry. A second, more secretive move was to offer these letters to employees, but under the letterhead "Crane Bros." An employee who was *not* involved in the strike would be provided with a letter with a special watermark that when held up to the light would show a crane (bird) standing erect; an employee who *was* a member of the strike would receive the letterhead with the watermark of a crane with a broken neck, serving as a warning to other potential employers that this person was involved in the strike.

As Professor Melvyn Dubofsky of State University of New York stated, "The Erdman Act proved so unsatisfactory to the unions and the companies that in 1913 Congress replaced it with the Newland Act. Although the new legislation created a permanent Board of Arbitration and Conciliation, one that could offer its services to the disputants on it own initiative, it proved in practice as ineffective as its predecessor."[16] Thus, the private sector had a powerful hold on the labor force in the late 1880s.

During this time, federal employees were not allowed to join unions, and the federal government had no obligation to bargain with a union. In 1912, the Lloyd–La Follette Act was passed, allowing protection to federal employees and outlining the "just cause" standard. The Lloyd–La Follette Act stated that "the right of employees, individually or collectively, to petition Congress or a member of Congress, or to furnish information to either House of Congress, or to a committee, or Member thereof, may not be interfered with or denied."[17] This was in response to the gag orders that Presidents Theodore Roosevelt (1902) and William Howard Taft (1909) had issued, prohibiting federal employees from communicating with Congress.

The only part missing from the Lloyd–La Follette Act was the enforcement section that would have allowed federal employees a means by which to take a claim in front of a court of law or any other administrative board. From this came the development of the Civil Service Reform Act of 1978 and codification in 5 *U.S. Code* § 7211. The Civil Service Reform Act was crafted with the idea that federal employees had a means of reporting unfavorable working conditions, acts of mismanagement, and abuse to Congress.[18]

In 1905, Mother Jones (fig. 2–4) helped form the Industrial Workers of the World (IWW), or "Wobblies." The IWW did not agree with the mission of the AFL. Membership in the organization was open (i.e., granted regardless of gender, race, and religion). The IWW was known for general strikes, boycotts, and acts of sabotage. The IWW was subjected to raids by the U.S. Department of Justice in September 1917 because of the organization's stance on the war. In connection with labor disputes, 165 IWW officers were arrested, more than 100 of whom were then convicted and sentenced to jail terms for violating the Espionage Act.[19] One of the biggest differences separating the Knights of Labor, the IWW, and the AFL was the type of organizational structure they formed. On the one hand, the Knights of Labor and the IWW approached issues as a single large union; in other words, no matter what area you came from, the union spoke with one collective voice. On the other hand, the AFL's approach was that each craft or trade has its own working conditions, and each should let the other define what is in the best interest of the given trade or craft.

Fig. 2–4. American labor activist Mother Jones (1837–1930). According to Clarence Darrow, she was "one of the most forceful and picturesque figures of the American labor movement."[20]

On March 15, 1917, the United States Supreme Court approved the Eight-Hour Act under threat of a national railway strike. In 1926, the labor movement would make gains through the passage of the Railway Labor Act. Strangely, labor and management both supported the Railway Labor Act, which would prevent the interruption of commerce. The act required railway owners to negotiate with employee unions in addition to providing for a means to resolve employer and employee differences through voluntary submissions to arbitration. It is widely considered to be the first comprehensive collective bargaining law.[21]

The CIO and the AFL

In 1935, the Congress of Industrial Organization (CIO) was formed with the efforts of the Amalgamated Clothing Workers of America. Interestingly, John L. Lewis, the founder of the CIO, was vice-president of the AFL until he resigned in November 1935. The membership roles for the CIO were based on status as industrial workers, not solely on the craft of a given worker. It was thought that the employees not having a given skill also needed representation in the workplace. John Lewis ended up in a long personal dispute with William Green about the leadership of the AFL and the campaign for industrial unionism. At the time, the AFL was not interested in industrial unionism.

The CIO would not be held back by the AFL's lack of interest; it would move to organize new unions in steel, glass, rubber, and automobiles. Late in 1937, the CIO estimated membership to be 3,700,000, while the estimated membership of the AFL was 3,400,000.[22] One significant difference between the two organizations was that the CIO championed for African Americans, women, and immigrants in the workplace.[23] It was not uncommon for a company to have the AFL represent one group of employees while the CIO represented another. Over the next few years, the distance between the AFL and the CIO grew, and each feared that the other would not properly represent its given direction.

The CIO, in its first show of strength, conducted the great sit-down strike against Chevrolet in Flint, Michigan. General Motors (GM) took its issue to court and was awarded a court order setting February 3, 1936, at 3:00 p.m. as the deadline for the union members to evacuate the building or face imprisonment. The CIO stood its ground and would not leave the building. The issue at hand was GM's unwillingness to enter into collective bargaining with the CIO. The disagreement was later settled, with GM agreeing to bargain with the CIO and not discriminate against union members. This was the first of many bold moves by the CIO. Unskilled workers were joining the CIO at increasing rates, and the AFL could see a firestorm before their eyes.

Then the 1937–38 economic downturn hit, which some referred to as the "Roosevelt depression." As a result, the CIO lost many dues-paying members, and eventually the AFL ended up with more dues-paying members than the CIO. The glory days of growth were just a memory.

In 1955, the two organizations merged as the AFL-CIO. Labor saw this as a new beginning, ushering in an era of renewed growth for labor.[24] This optimistic vision was not realized over the next decades: 25% of the labor force was unionized in 1955, as compared to 12.5% in 2004.

Now, more than 50 years after the AFL-CIO merger, several large unions have been leaving the group. July 25, 2005, the opening day of the 50th Annual Convention of the AFL-CIO in Chicago, is a day that many in labor will not forget for a long time, as a shockwave was sent through the labor movement when two unions filed for divorce from the AFL-CIO. The departure of the Service Employees International Union (SEIU), representing 1.7 million members (formerly the largest union in the AFL-CIO), and the International Brotherhood of Teamsters, representing 1.3 million members, reverberated with an enormous loss in membership numbers and a loss in annual dues of $20 million.[25] Further, three days later, the United Food and Commercial Workers, representing 1 million members, joined the SEIU and the Teamsters in announcing plans to disaffiliate from the AFL-CIO. Together, the disaffiliated unions had previously filed to create a new federation called the Change to Win coalition (CTW).

CTW was in disagreement as to how the AFL-CIO should move forward and how they are currently being administered. Andy Sterns, then leader of the SEIU, proposed three changes to the AFL-CIO:

- Spend more money organizing the unorganized workers, with at least 50% of AFL-CIO dues reimbursed for organizing.
- Streamline the AFL-CIO structure: the 56 unions would be reduced to 13 mega-unions within defined industries, thereby also eliminating jurisdictional disputes between and among unions.
- Organize global unions to challenge global capital and multinational corporations.

The AFL-CIO labor leadership offered a counterproposal; in the end, however, the AFL-CIO wanted nothing to do with the three proposed changes because many of its members felt they would be losing power in the organization. As a result of the defeat, Andy Sterns led the charge out of the AFL-CIO. At the time, many people did not know how this would end; some speculated this was possibly the final blow to labor, while others saw it as a productive move for the labor movement and long overdue. Almost immediately, UNITE HERE (a merger of the Union of Needletrades, Industrial, and Textile Employees and the Hotel Employees and Restaurant Employees International Union), Laborers Union, United Farm Workers of America, and the Carpenters Union all joined the CTW. As more unions joined the CTW, Andy Stern maintained that their goal was "not to divide the labor movement but to rebuild it."[26] It is still too early to condemn or condone the works of the CTW, and scholars in the field are keeping an eye on how it is affecting workers in the United States.

Fair Labor Standards Act

In 1938 the country witnessed the passage of the Fair Labor Standards Act (FLSA). The labor movement had been fighting for legislation to reduce the number of hours per day that an employee was required to work. Congress originally passed the FLSA to set the minimum wage for workers, restrict the maximum hours that a worker can work, define overtime requirements, and address child labor. Any violation of the FLSA is enforced by the U.S. Department of Labor.

Some would argue that this fight began over a century earlier by labor in an effort to move toward a 10-hour day. The FLSA trimmed the established 44-hour workweek to a 40-hour workweek over three years and took the 25-cents-per-hour wage and increased it to 40 cents per hour in seven years; an important last-minute amendment to the bill also prohibited labor by children under the age of 16. Over the years, the bill has gone through many amendments and has had a great impact on the fire service.[27] On May 14, 1947, the FLSA was amended by the Portal-to-Portal Act, which resolved the issue of what constitutes compensable hours worked under the FLSA. In 1949, the minimum wage was raised from 40 cents an hour to 75 cents an hour for all workers, and minimum wage coverage was expanded to include workers in the air transport industry. In 1961, the FLSA was again amended to expand the scope of coverage in the trade sector and to increase the minimum wage again (first to $1.15 and then to $1.25 in September 1963).

According to the U.S. Department of Labor, in 1974 Congress included under the FLSA all nonsupervisory employees of federal, state, and local governments and many domestic workers. (Subsequently, in 1976 in *National League of Cities v. Usery*, 426 U.S. 833, the Supreme Court held that the minimum wage and overtime provisions of the FLSA could *not* constitutionally apply to state and local government employees engaged in traditional government functions.) However, in a reversal, the Supreme Court in 1986 revisited the issue of the FLSA as it applied to public employees in *Garcia v. San Antonio Metropolitan Transit Authority*, 469 U.S. 528. Congress passed amendments changing the application of the FLSA to public sector employees. Specifically, these amendments permit state and local governments to compensate their employees for overtime hours worked with compensatory time off in lieu of overtime pay, at a rate of 1.5 hours for each hour of overtime worked. The 2007 amendments increased the minimum wage to $5.85 per hour effective July 24, 2007; $6.55 per hour effective July 24, 2008; and $7.25 per hour effective July 24, 2009.[28]

In the past 25 years, a number of cases have helped define the impact that the FLSA has on the fire service. In many cases, firefighters have unconventional work shifts compared to the private sector; for example, being allowed to rest (sleep) between runs and being allowed to eat while on duty have been the subject of many cases. To complicate this even more, the FLSA does not define full-time, part-time, or temporary employees. Therefore, the FLSA applies to all fire departments that employ firefighters. However, a person who volunteers their services to the fire department is not covered by the FLSA. Even if volunteers receive paid expenses or reasonable benefits, they still do not fall under the FLSA.[29] Section 207(a) of the FLSA requires employers to pay their employees overtime compensation for all hours worked more than 40 hours per week. However, law enforcement activities or fire protection activities have been specifically identified as "alternative 7(k) work periods," thus changing the normal 40-hour workweek/7-day period to a work period ranging from 7 to 28 days. Additionally, paramedics pose a challenge because they also perform firefighting duties as a normal job function, referred to as dual function.

The FLSA includes rules for the following topics: minimum wage, overtime, child labor, and the amount of hours that can be worked. In addition, many states have adopted specific wage and hour laws granting more rights than the FLSA. It is recommended that you review your particular state laws pertaining to wages and hours worked in addition to the FLSA. It is always recommended that you seek the advice of a legal expert on FLSA before taking any action.

Chronology of Significant U.S. Labor Relations and Fire Service Events

1678 Boston purchases the first motorized fire apparatus and pays someone to man the apparatus.

1790s First craft unions are formed (printers, shoemakers, tailors, and carpenters).

1790 Cabinet- and chair-makers in Philadelphia fight an attempt by employers to blacklist union members.

1791 Philadelphia carpenters strike unsuccessfully for a 10-hour day and overtime pay.

1806 Philadelphia Cordwainers case.

1818 First female volunteer firefighter hired.

1818 Federation of Organized Trades and Labor union formed (later dissolved to become the AFL in 1886).

1836 Massachusetts passes the first labor law to cover child labor.

1847 New Hampshire becomes the first state to pass a 10-hour workday.

1864 Contract Labor Law is passed, encouraging immigration (repealed 1868).

1869 Knights of Labor formed as a national union (farmers, skilled workers, and unskilled workers).

1875 The Molly Maguires case involving union members leads to strike-related violence.

1877 More than 100 workers killed in the national railway strike.

1881 Federation of Trade and Labor Union organized (later to become the AFL).

1886 Samuel Gompers organizes the first national trade union, the AFL.

1886 Haymarket Square riot results in the deaths of several workers, with blame for the violence going to the Knights of Labor.

1888 First federal labor relations law is passed, applying only to rail companies.

1890 The Sherman Antitrust Act is passed to break up corporate monopolies.

1892 Steelworkers go on strike in Homestead, PA.

1894 Pullman strike: railway workers go on strike over wages and rent issues.

1898 Erdman Act is passed (later declared unconstitutional by the U.S. Supreme Court in 1908).

1900 Washington, D.C., firefighters join the AFL union.

1905 The IWW (Wobblies) organizes to advocate one united voice for workers in a single union.

1914 The Clayton Act is passed, establishing limits to court injunctions, viewed as a major victory for labor.

1912 The Lloyd–La Follette Act is passed, extending protection to federal employees and outlining the just-cause standard.

1918 IAFF is formed, with the assistance of Samuel Gompers.

1929 The Railway Labor Act is passed, allowing railway workers to bargain collectively with owners and the use of voluntary arbitration.

1931 The Davis-Bacon Act is passed, requiring federal contractors to pay prevailing wages.

1932 The Norris–La Guardia Act establishes limitations to court injunction and voids the enforcement of yellow-dog contracts.

1935 The Wagner Act is passed, creating the National Labor Relations Board (NLRB).

1935 John Lewis leads the fight to break from the AFL and organizes the CIO.

1936 The United Automobile Workers (UAW) conduct the first sit-down strike, in Flint, Michigan.

1938 The FLSA is passed, allowing for a 40-hour workweek, overtime, and minimum wage, and creating a child labor law

1947 The Taft-Hartley Amendments are passed, resulting in amendment of the Wagner Act to balance the power between labor and management, as well as creating the Federal Mediation and Conciliation Service and the Right to Work states.

1950 Richmond, Virginia, hires its first African American firefighter.

1955 The AFL and CIO merge in hopes of providing a stronger voice for labor nationally, with George Meany as president.

1959 The Landrum-Griffin Act is passed to control the internal affairs of unions and their individual members.

1962 Executive Order 10988 is issued by President Kennedy, allowing federal employees to organize and enter into collective bargaining (individual states are also encouraged to do the same for state and local public sector employees).

1963 The Equal Pay Act is enacted.

1964 The Civil Rights Act is enacted, prohibiting discrimination on the basis of race or sex.

1968 The UAW leaves the AFL-CIO and joins the Teamsters in forming the Alliance for Labor Action.

1970 The Occupational Health and Safety Act is enacted to ensure a safe and healthy work environment for workers.

1970 Hawaii becomes the first state to allow state and local officials the right to strike.

1975 First legal large-scale public sector strike, by the American Federation of State, County and Municipal Employees, with 80,000 members.

1978 The Civil Service Reform Act is enacted.

1980 Joyce Miller becomes first woman to be appointed to the AFL-CIO executive board.

1981 In responding to the Professional Air Traffic Controllers Organization (PATCO) strike, President Reagan orders workers back on the job and permanently replaces workers refusing to return to work, a significant blow to labor and the use of strikes.

1988 The Workers Adjustment and Retraining Notification Act is enacted, requiring that companies with 100 or more workers must give 60 days advance notice of a shop closing.

1991 The Americans with Disabilities Act is enacted, protecting disabled persons from discrimination, amending Title VII of the Civil Rights Act.

1993 The Family and Medical Leave Act is enacted.

2005 The AFL-CIO splits, with seven major unions leaving the AFL-CIO and reorganizing as CTW.

2012 The IAFF is still growing and working on behalf of U.S. and Canadian firefighters' rights, in particular, collective bargaining, staffing, line-of-duty deaths, health care, safety, and pension issues.

Current Issues

Public sector collective bargaining underwent considerable growth during the 1960s and 1970s. In response to the increase in public sector collective bargaining, many states passed statutes that protected workers who joined labor organizations and who elected to enter into collective bargaining with a city or town. According to Gary Dessler, "With at least seven million public-sector union members, the public sector represents at least 44% of the total U.S. union membership. Three public unions—the National Education Association, the American Federation of State, County and Municipal Employees, and the American Federation of Teachers—are among the largest nine U.S. unions."[30]

Unions in the public sector today hold a considerable amount of political power. On the national front, in early 2008 the AFL-CIO had pledged to spend $53 million to send a Democrat to the White House.[31] Public unions usually employ full-time lobbyists whose sole duty is to lobby for labor issues. Many of the unions come out in support of a certain candidate in hopes that their candidate is elected and that they will benefit down the road when pro-labor issues are posed while that politician remains in office. However, these unions may be required to sit across from this same politician at the bargaining table, making for a strange relationship. Given the number of employees represented by the public union, it is clear why politicians often listen attentively to union concerns.

As states began developing statutes to address collective bargaining in the public arena, they frequently looked to the private sector to see what did and did not work. Many public sector statutes mirror the private sector and generally do not include strike clauses. One significant factor that makes bargaining different in the public sector is the heavy influence of politics. Harry Wellington and Ralph Winters said it best: "Collective bargaining by public employees and the political process cannot be separated. The costs of such bargaining, therefore, cannot be fully measured without taking into account the impact on the allocation of political power in the typical [jurisdiction]."[32] This topic is further discussed in chapter 3.

Labor relations in the United States involves the input of the government, with the creation of laws and regulations. Some will look at a labor law and say it favors management, and other laws may be seen to favor labor. Regardless, the creation of each law, act, regulation, or ordinance has a history. The goal of the government has been to encourage (and, in some cases, force) labor and management to work out their differences. Some firefighters and fire chiefs think that the union is all about conflict and that labor relations must be adversarial. This could not be further from the truth. Labor relations is about resolving differences within a set framework (grievances and arbitration) in a timely fashion, with open and honest communications and respect for all involved.

Many firefighters and administrators tend to lose sight of the history of the labor movement and get bogged down with only today's issues. Time and again we hear just how bad we have it—ignoring just how bad our early labor movement brothers and sisters truly had it. Let's face it: the fire service has not seen lives lost over labor issues/actions, unlike the early labor movement. Firefighters have not taken the fight to the streets or been killed in their pursuit of a better working environment, even though, unfortunately, firefighters have been killed by unsafe equipment and/or work environments over the years. Many firefighters have learned from the past sacrifices our brothers and sisters have made to better the service we are providing today.

One of the first recorded labor strikes for firefighters in the United States was in Boston, Massachusetts, in 1823. Firefighters take great pride in their occupation and have not seen the acts of strike violence that are often seen in others areas. However, firefighter labor unions have also been associated with acts of vandalism and numerous acts of arson during labor strikes across the United States. In 1978, Memphis, Tennessee, firefighters reached an impasse at the bargaining table with the city, and the firefighters went on strike. During the Memphis strike, the city was faced with vandalism and arson fires.[33] In 1993, the South Carolina Forestry Commission found that 33 fire department volunteers had been charged with arson, and the following year 47 firefighters were charged.[34] In May 2010, when the Mayor of Flint, Michigan, announced that 23 firefighters would be laid off, over the next 20 hours the city was faced with a rash of fires in an area of 30 miles.[35] One does not have to look far to see how some labor organizations have dealt with difficult labor relations. (Nynex, Greyhound, and Eastern Airlines have all been subjected to incidents of sabotage during difficult labor relations.)

According to the U.S. Department of Labor, in 2008, total paid employment in firefighting occupations was about 365,600. Firefighters held about 310,400 jobs; first-line supervisors/managers of firefighting and prevention workers held about 55,200; and fire inspectors and investigators held about 14,000 jobs. These employment figures include only paid career fire fighters; they do not cover volunteer firefighters, who perform the same duties.

As a result of legislation and court rulings, policies and procedures have been established in a myriad of employment areas to improve conditions or for general safety. In particular, the following areas have received focus:

- Occupational safety and health
- Equal employment opportunity
- Wages
- Health care
- Pensions
- Family leave

Labor and management should pay close attention to these areas, as they will most often be the issues at stake at the bargaining table and in the daily administration of the collective bargaining agreement. The increase in pension and health care costs to local governments will be areas of major concern for labor unions over the next several years.

Political lobbying can be accomplished either by the union side of the fire service or by nonunion fire service organizations. On the nonunion political front, organizations such as the National Volunteer Firefighters Council (NVFC) have a political agenda that supports the Assistance to Firefighters Grant Program, which provides funding directly to fire departments to purchase apparatus, training, and equipment. The NVFC supports the passage of the Volunteer Responder Incentive Protection Act, which would prevent the Internal Revenue Service from taxing the compensation that local volunteer firefighters and emergency medical responders receive for service provided to a community. The NVFC is pushing Congress to examine why the Department of Justice has been so slow in awarding benefits and why they have such a low rate of approval for benefits under the Hometown Heroes Survivors Benefits Act implementation (2003). The Department of Justice approved the final rules for determining claims under the act on September 11, 2006. As of July 13, 2007, there were 7 awards approved, 47 denials issued, and 199 cases still pending. Some 16 months dater, in December 2008, the DOJ approved 176 claims, 100 claims have been denied (32 of which are under appeal), and 76 remain pending. The NVFC has provided political pressure on the DOJ to ensure that claims are reviewed more quickly so that families in need are taken care of, which was the original intent of the Act.

The IAFF on the national front is working on Social Security issues affecting firefighters. Over the years it has lobbied individual states to enact presumptive disability laws; as of today, more than 40 states currently have such laws on the books. The presumptive laws cover conditions contracted on the job or that are considered job-related, such as certain cardiovascular diseases, cancers, and infectious diseases. These conditions are thus to be considered covered under state workers' disability compensation unless proved unrelated to an individual's occupation.

The IAFF is also lobbying for the passage of the Public Safety Employer-Employee Cooperation Act. According to the IAFF, Congress has expanded the scope of collective bargaining laws to protect private sector employees, nonprofit association employees, transportation workers, federal government employees, and, most recently, congressional employees. One of the few groups of workers not covered by federal law are state and local government employees, including public safety officers. Historically, Congress has taken a hands-off approach to telling local states how to manage their employees.[36] Currently, with the responsibility for homeland security being pushed back on the individual states and localities, the federal government should ensure that collective bargaining rights are offered to all. Many states today still enjoy the freedom of not allowing basic collective bargaining rights to the public sector.

Regardless of type or size, a fire and emergency medical service organization can have a significant impact on the labor relations process. Although I have chosen to focus on two of the larger political lobbying associations for the fire service, I do not mean to detract from the hard work and accomplishments at the state or even the local level. My point is that the labor relations process often involves politics. Later the book discusses how local politics can play a significant role in the collective bargaining process.

In modern times, one has to think back only a few years to the debate about the creation of the Department of Homeland Security and the Transportation Security Administration. The main focus of the labor-management debate was whether employees of the department had a right to join unions and bargain collectively. The Bush administration held its ground that the employees should not belong to a union because of the need for management to remain flexible in assigning jobs and job requirements.[37] On a national front, many unions saw this as a significant blow to labor. On the federal front, this shock wave was the most pronounced in 1981 with the strike by the former PATCO and the Federal Aviation Association, when nearly 13,000 employees walked off the job. PATCO thought they were untouchable. President Ronald Reagan ordered the workers back on the job by a specific date. Some members returned to the job, but many did not. In the end, 11,000 union employees received pink slips.[38] Replacement workers took the jobs of the PATCO members who were fired, and labor has been on the retreat ever since. Labor's long-standing power to strike had been put to the test and failed, representing to many unions the loss of a powerful tool from their political tool chest. Some have reported that the PATCO stance was about legislative issues—to achieve "wage determination on a private sector model."[39] One author claims that the standoff was over work-related problems, improving retirement issues, and addressing the autocratic management style.[40] Regardless of what the main focus was, PATCO took a strike vote and would not move off its mark, ultimately losing big in the end. PATCO did not fully understand the strength that the government would have and the lack of support they would receive from the general public. Many variables come into play when dealing with economic strikes. As any union member who has lived through a strike vote will tell you, it is not an enjoyable process.

On August 20, 2005 (less than a month after the AFL-CIO split), the Aircraft Mechanics Fraternal Association (AMFA) went out on strike against Northwest Airlines. Experts in the field have labeled this strike "PATCO II," because labor lost significantly. The AMFA was standing its ground and resisting concessions; however, the AFL-CIO and the CTW, in representing airlines workers, did not stand with the AMFA, and this caused the strike to fail. The lack of solidarity resulted in the AMFA's accepting major concessions and, as a further consequence, in Northwest Airlines' seeking major concessions from other unions. The ripple effects of this blow to labor could be seen near and far. So once again, history repeats itself and current labor organizations will learn from the past in order to move forward in a productive manner. We discuss this issue and the actions public sector employees can take later in this book.

Today labor is still fighting to provide stronger employee rights to the workers of the United States. Employers still maintain that the current labor laws are holding them back from free enterprise. Regardless of what side you are on, neither can agree that a perfect balance has been achieved. Labor organizations should be looking at how they envision themselves down the road. Will they take a back seat to the labor movement? Has the labor movement outlived its usefulness? No matter the issue that the labor movement faces, they need to collectively sit down and redefine the strategies and structures to respond in a positive manner to these issues.

This chapter has explored how, over the years, the labor movement has shifted with the ebb and flow of the business world, for good or bad. Work as we once knew it no longer exists. Technology, globalization, and demographics have all played a role in the "new work" we all face. Labor and management must adapt to these changing times. Political factors have always played a role in labor relations, and they will continue to do so. This cannot be more evident than with respect to public sector labor relations. As each state is free to establish the rules that they deem important at the time, political leaders will be involved. The public sector laws may also flow with the changing political climate.

Final Thoughts

Many would agree that the labor movement foundation in the U.S. is vested in the fight by workers on behalf of workers. Many of our ancestors may have been the ones fighting for better wages, better working conditions, and laws protecting the worker. Although our economy was created with the support of slavery, maltreatment, and inhumane working conditions, things have changed and the government has seen the immoral and unethical havoc it has caused or allowed in the past. Laws have been implemented in an effort to help employees and employers. What one thinks of a particular law depends on where one stands. Many of the labor laws that we benefit from today have come at the expense of lives before us. The fire service is fortunate not to have experienced the hundreds of losses in the streets that the founders of the labor movement faced, such as during the Railway Strike and Haymarket Square riots.

As we discussed at the beginning of this chapter, labor relations is affected by social issues, how employees and employers view the work environment, various classes, economics, and power. These issues came into play in many of the significant labor developments over the years.

Unions have witnessed highs and lows of membership, and at the same time, management has been required to produce more goods at a cheaper price to stay competitive. In the fire service, our own leaders have been pushed to improve the quality of our service, even while the demand for fire and emergency medical services has increased. Unions and local governments have all been faced with increasing pressure on how best to provide fire and emergency medical services to the community while maintaining a productive relationship.

It does not matter if you are union or nonunion, the collective force of the fire service should be about how best to protect our firefighters first and foremost. This chapter was meant to provide a solid foundation to build on in the next chapters. One must have a clear understanding and appreciation of the events that have led to our current labor relations system, and the only way to do this is to examine the historical events that have shaped it.

Key Terms

- American Federation of Labor (AFL)
- Civil Service Reform Act
- Clayton Act
- Cordwainers case
- Congress of Industrial Organization (CIO)
- Contract Labor Law
- Erdman Act
- Fair Labor Standards Act (FLSA)
- Federation of Organized Trades and Labor Unions
- Haymarket Riot
- Homestead and Pullman strikes
- International Association of Fire Fighters (IAFF)
- International Workers of the World
- Lloyd-LaFollette Act
- Knights of Labor
- Molly Maguires
- National Labor Relations Act
- Public Safety Employer-Employee Cooperation Act
- Railway Labor Act
- UNITE HERE
- Wagner Act

Exploring Issues on the Web

American Labor Studies Center: http://www.labor-studies.org/ (Collecting, analyzing, evaluating, and creating labor history and labor studies curricula and related materials, aligned to the various state and national standards and disseminated to kindergarten through 12th-grade teachers nationwide)

George Meany Memorial Archives Online: http://www.georgemeany.org/archives/home.html (Providing reference assistance and more, at the archives established by the AFL-CIO to preserve its historical records and make them available for research)

Industrial Relations Library (IR): http://www.princeton.edu/~ssrc/IRmain.html (Organized in 1922, coinciding with the establishment of Princeton University's Industrial Relations Section, IR ranks as the oldest special collection of its type among universities in the United States. The collection of 6,500 volumes and over 100,000 pamphlets is highly specialized.)

Labor and Working Class History Association (LAWCHA): http://www.lawcha.org/ (Open to everyone interested in studying the history of the working class, their lives, workplaces, communities, organizations, cultures, political activities, and societal contexts)

Samuel Gompers Papers Online: http://www.history.umd.edu/Gompers/index.htm (Documentary history of the American working class, sponsored by the University of Maryland–College Park, the National Historical Publications and Records Commission, the National Endowment for the Humanities, and the George Meany Memorial Archives)

Tamiment Library and Robert F. Wagner Labor Archives at New York University: http://www.nyu.edu/library/bobst/research/tam/index.html (Unique, internationally known center for scholarly research on labor and the Left. The primary focus is the complex relationship between trade unionism and progressive politics and how this evolved over time.)

U.S. Department of Labor, Bureau of Labor Statistics, Occupational Outlook Handbook, 2008-09 Edition, Fire Fighting Occupations: http://www.bls.gov/oco/ocos158.htm

U.S. Department of Labor Library: http://www.dol.gov./oasam/library/ (One of the oldest Cabinet-level libraries, established in 1917. The online catalog provides access to all materials that have been acquired by the library since 1975, as well as access to selected pre-1975 items. Not all catalog records have been converted to electronic format, however, and a card catalog still provides access to most holdings acquired prior to 1975.)

U.S. Labor and Industrial History World Wide Web Audio Archive, Department of History, University at Albany, State University of New York: http://www.albany.edu/history/LaborAudio/

Voices of Labor Oral History Project: http://www.library.gsu.edu/spcoll/Labor/voices/index.htm (Recording the personal experiences of men and women who participated in the labor movement)

References

1. Thompson, B. 1988. The evolution of fire services. In *Managing Fire Services*, ed. R. J. Coleman and J. A. Granito. 2nd ed. Washington, D.C.: International City/County Management Association. P. 24.
2. San Francisco. 2007. The early years of the department. http://www.sfmuseum.org/hist1/hist2.html (accessed January 25, 2008).
3. Crawford, R. 2007. Journal of San Diego history. San Diego Historical Society. http://www.sandiegohistory.org/journal/89spring/eating.htm (accessed May 11, 2007).
4. Holley, W., K. Jennings, and R. Wolters. 2005. *The Labor Relations Process*. 8th ed. Mason, Ohio: South-West. P. 85.
5. Carter, H., and E. Rausch. 1994. *Management in the Fire Service*. 2nd ed. Quincy, Mass.: National Fire Protection Association. P. 28.
6. Bakersfield, California. 2007. BFD historical timeline. http://www.ci.bakersfield.ca.us/FIRE/departprofile/HistoricalTimeline.htm (accessed January 2008).
7. Dulles, F., and M. Dubofsky. 1993. *Labor in America*. Wheelin, Ill.: Harlan Davidson. P. 92.
8. AfroAmerican Web Ring (AAWR). n.d. History of black firefighters. http://hometown.aol.com/fireriter (accessed May 15, 2007).
9. Gorn, E. 2001. Mother Jones: The most dangerous woman in America. New York: Hill and Wang.
10. Dulles and Dubofsky 1993, p. 112.
11. Kenny, K. 1998. *Making Sense of the Molly Maguires*. New York: Oxford Press. P. 6.
12. Dulles and Dubofsky 1993, pp. 117–19.
13. Taylor, T. 2000. The Haymarket Square Riot. Weird and Haunted Chicago. http://www.prairieghosts.com/haymarket.html (accessed May 15, 2007).

14. Boller, H. R., and D. J. Petersen. 2006. The use of union insignia and emblems in the workplace. *Employee Relations Law Journal*. 14: 47.

15. Dubofsky, M. 1994. *The State and Labor in Modern America*. Chapel Hill, N.C.: University of North Carolina Press. P. 32.

16. Ibid., p. 58.

17. Lloyd–La Follette Act. August 24, 1912, § 6, 37 Stat. 555, 5 U.S.C. § 7511.

18. Starling, G. 1998. *Managing the Public Sector*. 5th ed. Orlando, Fla.: Harcourt Brace and Company. P. 492.

19. Goldman, E. 2004. People & Events: Prelude to the Red Scare: The Espionage and Sedition Acts. PBS Online. http://www.pbs.org/wgbh/amex/goldman/peopleevents/e_redscare.html (accessed September 2, 2010).

20. Ockerbloom, Mary (n.d.) Mary Harris (Mother) Jones. http://digital.library.upenn.edu/women/jones/MotherJones.html (accessed November 16, 2011).

21. Holley, W., and K. Jennings. 1997. *The Labor Relations Process*. 6th ed. Fort Worth, Tex.: Dryden Press. P. 108.

22. Dulles and Dubofsky 1993, pp. 279–88.

23. Carrell, M., and C. Heavrin. 2007. *Labor Relations and Collective Bargaining*. 8th ed. Upper Saddle River, N.J.: Pearson Prentice Hall. P. 18.

24. Dulles and Dubofsky 1993, pp. 289–300.

25. Edsall, T. B. 2005. Two top unions split from AFL-CIO. *Washington Post* http://www.washingtonpost.com/wp-dyn/content/article/2005/07/25/AR2005072500251.html (accessed February 2008).

26. Light, R. O. 2005. The AFL-CIO split and the fight for workers' power. http://www.mltranslations.org/US/ROL/ROLaflcio.htm (accessed January 2008).

27. Dulles and Dubofsky 1993, p. 274.

28. U.S. Department of Labor. 2008. U.S. Department of Labor history of changes to the minimum wage law. http://www.dol.gov/esa/minwage/coverage.htm (accessed April 1, 2008).

29. Varone, J. C. 2007. *Legal Considerations for Fire and Emergency Services*. Clifton Park, N.Y.: Thomson Delmar Learning. Pp. 457–58.

30. Dessler, G. 2008. *Human Resource Management*. 11th ed. Upper Saddle River, N.J.: Pearson Prentice Hall. P. 624.

31. Mullins, B. 2008. AFL-CIO targets McCain with $53 million campaign. *Wall Street Journal*. http://blogs.wsj.com/washwire/2008/03/11/afl-cio-targets-mccain-with-53-million-campaign/?mod=googlenews_wsj (accessed March 18, 2008).

32. Wellington, H., and R. Winter. 1971. *The Unions and the Cities*. Washington, D.C.: Brookings Institution. P. 31.

33. Thieblot, A., and T. Haggard. 1983. Memphis police and firefighter strike of 1978. http://www.ncjrs.gov/App/Publications/abstract.aspx?ID=123939 (accessed September 2, 2010).

34. U.S. Fire Administration. 2003. Special report: Firefighter arson (USFA-TR-141). http://www.usfa.dhs.gov/downloads/pdf/publications/tr-141.pdf (September 2, 2010).

35. Angus, L. 2010. Rash of fires in Flint thought to be for "perverted political purpose." http://www.mlive.com/news/flint/index.ssf/2010/03/rash_of_fires_in_flint_thought.html (accessed September 2, 2010).

36. IAFF. 2007. Our history: Honor, struggles, and action. http://www.iaff.org/about/history/ourhistory.htm (accessed May 11, 2007).

37. Harnage, B. L. 2002. Personnel issues in establishing the new department. Statement before the House Select Committee on Homeland Security. July 17. http://www.afge.org/ (accessed July 20, 2007).

38. Carrell and Heavrin 2007, p. 256.

39. Northrup, H. R. 1984. The rise and demise of PATCO. Industrial and Labor Relations Review. 37 (2): 167–184.

40. Hurd, R. W., and J. K. Kriesky. 1986. Communications: The rise and demise of PATCO reconstructed. Industrial and Labor Relations Review. 40 (1): 115–127.

3 LABOR-MANAGEMENT RELATIONS IN THE PUBLIC SECTOR

Learning Objectives

After reading this chapter, you should be able to:

- Compare and contrast public sector and private sector labor relations laws.
- Identify labor relations laws that affect public sector bargaining for public safety.
- Describe the impact that the no-strike rule has had on the labor relations process in the fire service.
- List current trends in union membership.

Labor-management relations are critical to the success of the fire department, regardless of whether you are union or nonunion. The issues discussed in this chapter apply universally to career, call, and volunteer fire service organizations in the public sector. Labor-management relations in the public sector is a distinct animal compared to labor-management relations in the private sector. Public sector labor relations is an outgrowth of the private sector labor relations field. In the public sector, the labor relations laws, ordinances, regulations, and executive order (EO) are crafted by individual states, rather than by a single federal law, such as the National Labor Relations Act (NLRA) or the Federal Labor Relations Act. In light of this significant difference between public and private sector labor relations, this chapter will directly address how public sector labor relations work. Rather than outline each state's specific laws, we will review some of the common labor laws and identify some of the individual state laws as they pertain to labor relations. The individual firefighter and fire service administrator should have a clear understanding of the rights and privileges pertaining to public sector relations. From a personnel management standpoint, labor laws will affect how you perform your duties and responsibilities.

The public sector is no different from the private sector when it comes to doing more for less. Public sector and private sector administrators have been faced with the challenge of doing business better at a lower cost. Unfortunately, workers are often at odds with management when it comes to doing more for less. Taxpayers are demanding more services for their tax money. Government waste and overspending are constantly under the microscope of taxpayers and government watchdog organizations.

Some labor-management programs have helped deliver a better service to the public—our customers—in a more efficient manner. Only a few years back, there was a major push to privatize many aspects of the public service. Today, we have seen a full-fledged push by many states to reduce or remove many of the collective bargaining laws that cover public sector unions. Many in the public sector saw privatization as a significant threat, and they mobilized resources to prevent this from happening. Because labor and management both had to work on ways to provide better and lower-cost service to the customers, they entered into a partnership. The following sections examine productive labor-management partnerships in local government.

Public sector unions today tend to include more professionals (teachers, professors, firefighters, and law enforcement) than did the older trade unions, often referred to as blue-collar workers in the 1960s and 1970s. The same can also be seen in the private sector, where workers of high-tech jobs and even doctors and nurses are being unionized. Today's unions are not your grandfathers' unions. Although the mind-set of the 1960s-era unions still resonates among labor, today's union members are often well educated and hold professional jobs.

The *labor relations process* describes when two parties come together to jointly determine and administer work rules. While many people automatically associate labor relations with labor unions, the labor relations process can be used in a union work environment, a nonunion work environment, or a volunteer organization. In the context of labor unions, the labor relations process can be broken down into three phases (fig. 3–1):

- *Recognition.* Defining the rights and responsibilities of labor and management.
- *Negotiation.* Negotiating a labor agreement and determining a dispute resolution process.
- *Administration.* Applying and enforcing the terms and conditions as outlined in the negotiated labor agreement.

Many of the issues discussed in this book will fall under one of these three steps. The labor relations process follows these three steps in sequence, each of which cannot exist without the others. As will become evident, some fire service agencies may have fallen short in the second step, while others only apply the third step sporadically. Often the labor relations process will build on each step over the years.

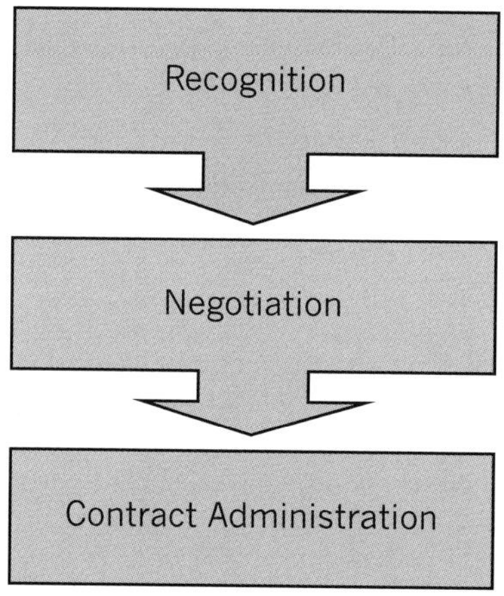

Fig. 3–1. Three phases of the labor relations process

For the purpose of this book, consider public sector collective bargaining as the employees' rights to organize a union and the right to enter into discussions over work conditions, work location, wages, and safety—hence the term *collective bargaining*. Importantly, the collective bargaining laws for each state are different for the public sector; however, they can generally be broken down into three sections (fig. 3–2):

- *Binding interest arbitration.* The employer and the union are required to bargain.
- *Meet and confer.* The employer and the union need only meet and confer over items; there is not a defined process for impasse resolutions.
- *Bargaining not required.* The state does not have any public sector bargaining laws or any statutorily required regulations. States without collective bargaining laws include Arizona, Alabama, Georgia, Mississippi, New Mexico, and Texas. Public sector employees' right to organize and belong to unions is settled in federal law, state laws, and many U.S. Supreme Courts decisions. What is missing in some states is the right to enter into a collective bargaining agreement. Therefore, firefighters could belong to a union and lack the legal standing to enter into a collective bargaining agreement with the government.

The following sections examine how local politics can play a role in the labor relations process. This can be so small that a member of the selectmen in a small town can halt years of progressive bargaining. We will look at the strength that public sector unions have on the national and local political fronts. We will examine how this can be either an important ingredient for a successful outcome or a liability for the labor relations process.

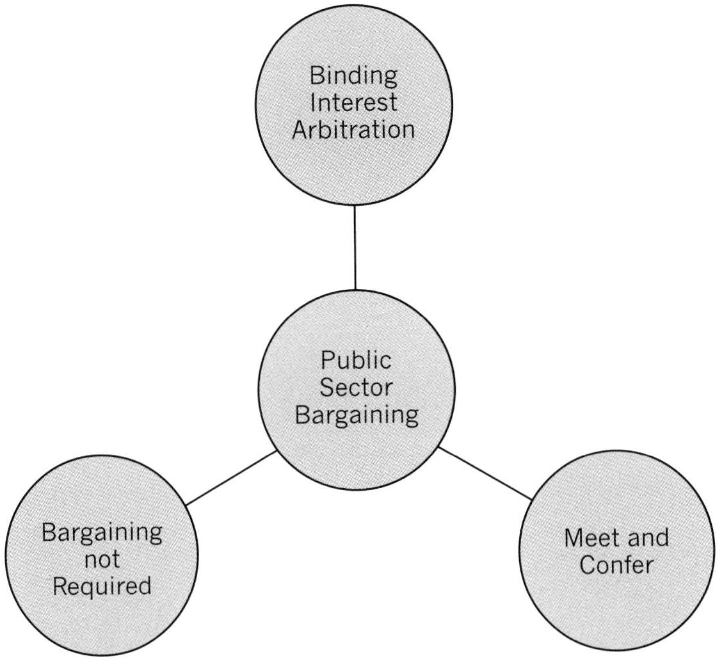

Fig. 3–2. Sections of public sector bargaining

Chapter 2 demonstrated how many of the labor relation laws have been enacted over the years, and here we will show that they play a role in the public sector labor relations process. The following sections examine some of the state laws on labor relations and the limitations and benefits from these laws. We will identify some significant labor relations cases involving firefighters across the United States.

Public Sector Labor Relations

The National Industrial Recovery Act of 1933 is considered the beginning of the private sector's ability to organize and bargain collectively through unions. The next significant improvement to the labor movement was the passage of the NLRA in 1935. By the end of World War II, the United States witnessed collective bargaining's growing acceptance in the private sector. At the time, the mining and manufacturing companies reluctantly entered into wage agreements with the unions. A labor-management relationship was born.

Before 1963, the U.S. Postal Service organized unions in an effort to lobby Congress. Public sector (federal) labor relations was born when President Kennedy issued EO 10988 (1962), allowing federal employees to organize and obligating management to recognize and enter into collective bargaining with employee organizations. Federal employee associations (unions) were nonexistent prior to

1962. Eight years later, it was estimated that about 65% of eligible federal employees had been organized. As a result of EO 10988, the nation witnessed the public sector's authorization to bargain collectively.

EO 11491 (October 1969), EO 11616 (August 1971), and EO 11838 (February 1975) extended the principles and objectives of EO 10988. The new order outlined the bargaining process for federal employees. In 1978, President Jimmy Carter crafted the Civil Service Reform Act, specifying that the president no longer has the authority to regulate collective bargaining on his own behalf.[1] Between the early 1960s and the late 1970s, public sector collective bargaining became law and developed into the detailed process we retain today.

The newborn public labor relations process seemed to work until the 1970s. The United States witnessed a decrease in labor relations across the board, and owners of companies began to rebuff the unions' demands. Some companies relocated to parts of the United States with lower wages and a cheaper work environment. Some 30 years later, as a result of globalization, the U.S. workforce is facing a similar dilemma: work that had been performed by union employees is being shipped to nonunion workers overseas. Because of globalization, the private sector is no longer bound to the shores of the United States; rather, labor relations can now stretch around the world. Today many unions are forced to accept major wage and benefit reductions in order for companies to remain operational and profitable in the private sector; if the union did not, the plant would send the jobs overseas, and the workers would have no jobs. Consequently, many unions are making significant concessions in order to keep jobs in the United States.

The fire service has been fortunate to be the only game in town—that is, fire departments are not in competition with other fire departments for work. The local community cannot send fire service responsibilities outside the community or overseas. Some have argued that this is a form of monopoly that has resulted in the fire service's resistance to change. A fire department does not have to show a profit-and-loss margin when presenting a business plan to the taxpayers. The average taxpayer does not know how to differentiate a good fire department from a bad fire department. The general public often lacks any solid quantifiable or qualitative data to assess the fire department's performance. The public will often rate the local fire department on the basis of the number of fires in the community or the cost of property loss due to fire. Because of this, some communities are budgeting good money on poor services.

Notably, the U.S. Congress did not include federal, state, and local government employees under the NLRA (fig. 3–3). It was not until 1970, with the passage of the Postal Reorganization Act, that postal employees were permitted to come under the NLRA. As Carrell and Heavrin have pointed out, Congress at the time viewed government workers as public servants and not employees; this viewpoint left many government workers without protection.[2] The end result of the patronage in government came at the cost of employment security. Government employees did not hold positions for a long time; instead, they held the positions based on who they liked, campaigned for, contributed money to, and so forth. In other words, if you supported the right political candidate, you might get to keep your job; if you supported the candidate who lost, you would most likely lose your job.

Fig. 3–3. Congress delegated the power to each individual state to enact its own labor relations laws for state and local public employees. (Photo courtesy of John Jenssen.)

Congress' view of government employees as servants indicates that the labor movement in the public sector will have to approach collective bargaining differently from the private sector. Many of the differences are based in labor policies and the legal and institutional framework when it comes to collective bargaining in the public sector. As outlined in chapter 2, many private sector labor policies were developed over the years, yet the public sector employee was exempted from these policies. Public sector labor polices are considered youthful compared to those in the private sector, and many public sector polices are the direct result of private sector policies. For example, many state collective bargaining regulations utilize the NLRB's outline for collective bargaining, arbitration, and impasse issues. The states of New York, Wisconsin, and Pennsylvania have passed legislation allowing public sector employees the same rights as those in the private sector.

The right to bargain is a good starting point to illustrate the difference between the private and public sectors. In the private sector, the extent and limits of collective bargaining have been clearly defined over time. Labor and management in the private sector must bargain collectively; however, it is not explicitly stated that they must agree to a proposal. Not agreeing to a proposal is called an *impasse*, which is discussed in detail in chapter 5. In the public sector, bargaining collectively can become clouded, depending on the state. One cannot find a standardized policy establishing collective bargaining in the public sector. Some states grant organizing and/or bargaining rights to public sector employees. In other states, they may require only that state and municipal employers "meet and confer" with employee organizations/representatives. Interestingly, meet and confer is often reduced to a nonblinding *memorandum of understanding* (MOU) between the two parties. Additionally, the meet-and-confer requirements do not have any means to resolve an MOU impasse. However, the following states do not currently have legislation allowing public sector employee to engage in collective bargaining:[3]

- Alabama
- Arizona
- Arkansas
- Colorado
- Louisiana
- Maryland
- Mississippi
- Missouri
- North Carolina
- South Carolina
- Virginia
- West Virginia

In North Carolina and Virginia, it is illegal for the state or a municipal entity to negotiate, collectively bargain with, or enter into an agreement with a public sector union.[4] Some states with legislation for collective bargaining view it as merely the requirement to meet and discuss issues with employees; the legislation is so weak that it does not bind the local government or state government to bargain collectively over wages, working conditions, and safety.

Code of Virginia

Title 40.1—Labor and Employment, § 40.1–57.2

Prohibition against Collective Bargaining

No state, county, municipal, or like governmental officer, agent or governing body is vested with or possesses any authority to recognize any labor union or other employee association as a bargaining agent of any public officers or employees, or to collectively bargain or enter into any collective bargaining contract with any such union or association or its agents with respect to any matter relating to them or their employment or service.

Source: http://leg1.state.va.us/cgi-bin/legp504.exe?000+cod+TOC4001000

State of Arizona

Title 23—Labor, Chapter 8, Article 6: 23-1411.

Public Safety Employee Organizational Rights Definition

A. Public safety employees serving any city, town, county, or fire district in this state have the right to join employee associations which comply with the laws of this state and have freedom to present proposals and testimony to the governing body of any city, town, county or fire district and their representatives. A person shall not be discharged, disciplined or discriminated against because of the exercise of these rights.

B. This section shall not be construed to compel or prohibit in any manner any employee wage and benefit negotiations.

C. For the purposes of this section, "public safety employee" includes a probation officer or surveillance officer who is employed by this state or a political subdivision of this state.

State of Nevada

Chapter 288—**Local Government Employee-Management Relations Act**

The following sections defines 21 *mandatory* items that are subject to bargaining:

Sick leave, vacation leave, holidays, other paid or nonpaid leaves of absence, insurance benefits, total hours of work required of an employee on each workday or workweek, total number of days' work required of an employee in a work year, discharge and disciplinary procedures, recognition clause, method used to classify employees in the bargaining unit, deduction of dues for the recognized employee organization, protection of employees in the bargaining unit from discrimination because of participation in recognized employee organizations consistent with the provisions of this chapter, no-strike provisions consistent with the provisions of this chapter, grievance and arbitration procedures for resolution of disputes relating to interpretation or application of collective bargaining agreements, general savings clauses, duration of collective bargaining agreements, safety of the employee, teacher preparation time, materials and supplies for classrooms, policies for the transfer and reassignment of teachers, and procedures for reduction in workforce.

Source: http://www.leg.state.nv.us/nrs/NRS-288.html

Wisconsin was one of the first states to pass a law requiring local government to bargain collectively with labor organizations. Yet, in 2011, Wisconsin set the stage for the new generation of public sector bargaining. Wisconsin passed a law to allow benefits to be discussed outside the constriction of the collective bargaining agreements, which had long been a protected right of unions in Wisconsin (fig. 3–4). According to the U.S. General Accounting Office (US-GAO), "26 states and the District of Columbia have laws that provide collective bargaining rights to essentially all public employees."[5] The US-GAO further reports that "12 states essentially do not have any laws for collective bargaining among state and local employees," and these states have granted specific groups (e.g., state workers, teachers, or firefighters) bargaining rights, but *not* to all state and local government workers.[6] For example, Texas allows police and fire services collective bargaining with approval from a majority of voters, but all other public collective bargaining is prohibited.

Florida and New Jersey are the only two states that have expressly granted collective bargaining in their state constitutions. In general, collective bargaining in the public sector is found in state statute, charter, ordinances, and EO.[7] Many states have created a state agency or board to address labor relations for the state, analogous to the role that the NLRB holds for the private sector.

Fig. 3–4. Public sector employees gathered at the State Capitol in Madison, Wisconsin, in protest of the proposed collective bargaining laws offered by Governor Scott Walker. (Photo courtesy of John Jenssen.)

Walter Gershenfeld and Joel Cutcher-Gershenfeld maintain that public sector laws governing collective bargaining are rooted in legislation covering the private sector. The NLRA and the Labor Management Relations Act provide the foundation for public sector labor relations. They outline seven traits that distinguish public sector and private sector bargaining:[8]

1. *Agency mission*. The public sector is not allowed by law to bargain the mission of the agency. The private sector has some leniency in setting the company mission with labor.
2. *Multilateralism*. Public sector bargaining often involves parties not sitting directly at the negotiation table. A collective bargaining agreement may be agreed to by the two parties but must be approved by the local voters. Public sector employees may seek pressure from politicians (legislation personnel, boards of selectmen, city aldermen) and from the community in presenting their issues. On the political aspects, these elected officials may need to lobby on behalf of the union in an effort to gain political support at election time. Let's face it: money and votes get people elected, and the union can be your friend or your enemy; in many states, it is the union that can add stability and electability to a person.
3. *Legislative support*. Many state and local collective bargaining agreements dealing with financial expenditures must be approved (voted) by the legislative body (taxpayers, finance committees, selectmen, or aldermen). These people may not be directly involved with the negotiations.
4. *Conflict with law*. Some state and local laws may be in conflict with provisions of the collective bargaining agreement. As a result of court

decisions and new legislation, the agreement must spell out how this will be handled and what laws have precedence.

5. *Professionals*. Public sector bargaining generally involves professionals such as teachers, nurses, and professors. Many public sector agreements reflect this in what is important to a given subject area. Some private sector unions are made up of mostly unskilled workers.
6. *Interest arbitration*. Currently, many public sector union contracts contain interest arbitration in the event that the two parties reach an impasse during contract negotiations. This also includes new contracts.
7. *Role of the strike*. Most public sector unions are prohibited from striking. Therefore, public unions must resort to other means to take a stand against a state or local agency.

The aforementioned, though not an all-inclusive list of differences for public sector bargaining, outline several major issues. Many states adopted public sector bargaining laws in the mid-1960s. So, public sector collective bargaining has not been around all that long, and some would argue that it is still developing. Over the past 20 years, the United States has not witnessed any new significant public sector collective bargaining legislation. In the 1980s, the states of Ohio and Indiana and, in 1992, New Mexico passed legislation dealing with collective bargaining.

Union membership

Private sector unions have seen a decrease in membership year after year, while the public sector has witnessed a steady increase in union membership. Because of the increasing number of public sector union members, it is imperative that today's fire service administrators have the knowledge and education to deal effectively with organized labor. In 2007, the U.S. Department of Labor's Bureau of Labor Statistics reported that the number of workers belonging to a union had risen by 311,000, to 15.7 million.[9] In 2009, the Bureau of Labor Statistics reported that the number of workers belonging to a union had declined to 15.3 million; altogether, 16.9 million wage and salary workers were represented by a union in 2009 (table 3–1), including union members (15.3 million) and workers who reported no union affiliation but whose jobs were covered by a union contract (1.6 million).[10] In 1983, the first year for which comparable union data were available, the union membership rate was 20.1%. The following are highlights from the 2009 data:[11]

- More public sector employees (7.9 million) belong to a union than private sector employees (7.4 million), despite there being five times more wage and salary workers in the private sector.
- Education, training, and library occupations had the highest unionization rate among all occupations, at 38.1%.
- The State of New York had the highest union membership rate, at 25.2%, compared to the State of North Carolina with the lowest membership rate, at 3.1%

Table 3–1. Union affiliation of employed wage and salary workers

NUMBER OF EMPLOYED	FEDERAL GOVERNMENT	STATE GOVERNMENT	STATE GOVERNMENT	PUBLIC GOVERNMENT
2000 REPRESENTED BY UNIONS	1186	1867	4923	7976
2008 REPRESENTED BY UNIONS	1167	2167	5342	8676
2009 REPRESENTED BY UNIONS	1192	2222	5263	8677

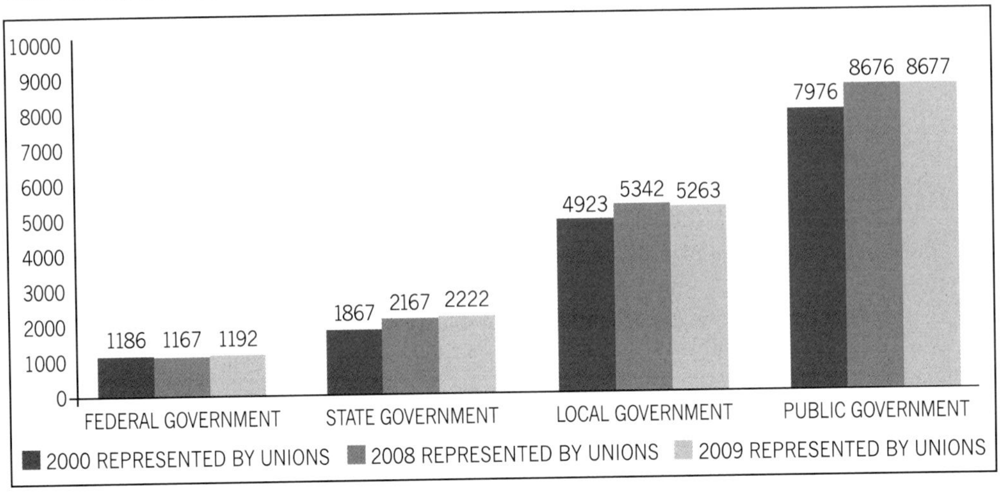

The U.S. Department of Labor reported that 13.2% of men belonged to unions, compared to 11.3% of women in 2009. Between 1983 and 2009, the union membership rate for men declined by 11.4 percentage points, and the rate for women declined by 3.3 percentage points. By age, the lowest membership was reported in individuals age 16–24 years, with a rate of membership of only 4.7%. The highest rate of membership was in workers age 55–64 years, with a rate of membership of 16.6%.[12]

Union membership continues to grow in the public sector, whereas in the private sector, there has been a decline in membership over the past 30 years (in 1980, 23% were union members; in 1995, 14.9% and in 2010, 12.3%).[13] Table 3–2 and figure 3–5 identify the number of union members from each state and people who are not members of the union but hold jobs covered by a union agreement. Therefore, the focus has been to educate and train public sector administrators and managers to deal effectively with the increase in public union membership.

The decline in union membership nationally may be attributed to three general workplace trends:

- Better laws/political conditions
- Changing workforce
- Better working conditions/administration

Table 3–2. Union affiliation of employed wage and salary workers by state (numbers in thousands)

State	2009					2010				
	Total employed	Members of unions[1]		Represented by unions[2]		Total employed	Members of unions[1]		Represented by unions[2]	
		Total	Percent of employed	Total	Percent of employed		Total	Percent of employed	Total	Percent of employed
Alabama	1,763	191	10.9	212	12.0	1,809	183	10.1	203	11.2
Alaska	293	65	22.3	69	23.6	295	68	22.9	73	24.8
Arizona	2,471	162	6.5	194	7.9	2,507	161	6.4	203	8.1
Arkansas	1,103	47	4.2	55	5.0	1,082	44	4.0	59	5.4
California	14,297	2,453	17.2	2,622	18.3	13,892	2,431	17.5	2,578	18.6
Colorado	2,175	153	7.0	181	8.3	2,130	140	6.6	171	8.0
Connecticut	1,538	265	17.3	282	18.4	1,549	258	16.7	270	17.4
Delaware	367	44	11.9	47	12.8	352	40	11.4	44	12.5
District of Columbia	276	29	10.4	35	12.5	287	26	9.0	30	10.5
Florida	7,097	411	5.8	489	6.9	7,033	392	5.6	488	6.9
Georgia	3,869	177	4.6	226	5.9	3,792	153	4.0	191	5.0
Hawaii	526	123	23.5	128	24.3	511	111	21.8	120	23.5
Idaho	577	36	6.3	46	7.9	584	42	7.1	50	8.6
Illinois	5,435	951	17.5	997	18.3	5,434	844	15.5	891	16.4
Indiana	2,612	277	10.6	319	12.2	2,554	279	10.9	313	12.2
Iowa	1,398	156	11.1	185	13.3	1,393	158	11.4	192	13.8
Kansas	1,249	77	6.2	104	8.4	1,222	84	6.8	111	9.1
Kentucky	1,657	142	8.6	173	10.5	1,642	147	8.9	166	10.1
Louisiana	1,704	99	5.8	110	6.5	1,742	76	4.3	96	5.5
Maine	543	63	11.7	74	13.7	543	63	11.6	71	13.0
Maryland	2,555	323	12.6	366	14.3	2,558	296	11.6	329	12.9
Massachusetts	2,864	476	16.6	516	18.0	2,866	415	14.5	446	15.6
Michigan	3,785	710	18.8	752	19.9	3,806	627	16.5	659	17.3
Minnesota	2,400	362	15.1	377	15.7	2,468	385	15.6	397	16.1
Mississippi	1,029	49	4.8	66	6.4	1,037	46	4.5	58	5.6
Missouri	2,481	234	9.4	264	10.6	2,469	244	9.9	274	11.1
Montana	374	52	13.9	68	18.1	364	46	12.7	52	14.4
Nebraska	823	76	9.2	94	11.4	813	75	9.3	96	11.8
Nevada	1,097	173	15.7	188	17.2	1,012	151	15.0	170	16.8
New Hampshire	616	67	10.8	76	12.3	622	63	10.2	73	11.7
New Jersey	3,734	721	19.3	742	19.9	3,734	637	17.1	660	17.7
New Mexico	759	51	6.7	77	10.2	750	55	7.3	72	9.7
New York	8,021	2,019	25.2	2,182	27.2	8,078	1,959	24.2	2,099	26.0
North Carolina	3,707	115	3.1	162	4.4	3,686	117	3.2	180	4.9
North Dakota	301	21	6.8	30	9.8	313	23	7.4	28	9.1
Ohio	4,827	685	14.2	742	15.4	4,787	655	13.7	702	14.7
Oklahoma	1,456	83	5.7	107	7.3	1,418	77	5.5	99	6.9
Oregon	1,471	250	17.0	272	18.5	1,515	245	16.2	268	17.7
Pennsylvania	5,220	782	15.0	844	16.2	5,224	770	14.7	831	15.9
Rhode Island	444	80	17.9	83	18.7	456	75	16.4	79	17.4
South Carolina	1,672	75	4.5	91	5.4	1,713	80	4.6	107	6.2
South Dakota	357	20	5.5	24	6.6	357	20	5.6	24	6.6
Tennessee	2,387	121	5.1	156	6.6	2,477	115	4.7	142	5.8
Texas	9,920	508	5.1	615	6.2	10,025	545	5.4	677	6.7
Utah	1,136	79	6.9	90	8.0	1,144	75	6.5	96	8.4
Vermont	285	35	12.3	40	14.1	289	34	11.8	40	13.6
Virginia	3,503	166	4.7	191	5.4	3,473	161	4.6	196	5.7
Washington	2,847	574	20.2	612	21.5	2,837	552	19.4	605	21.3
West Virginia	699	97	13.9	108	15.4	677	100	14.8	111	16.5
Wisconsin	2,528	385	15.2	400	15.8	2,508	355	14.2	380	15.1
Wyoming	239	18	7.7	20	8.3	246	18	7.4	21	8.4

1 Data refer to members of a labor union or an employee association similar to a union.

2 Data refer to both union members and workers who report no union affiliation but whose jobs are covered by a union or an employee association contract.

NOTE: Data refer to the sole or principal job of full- and part-time wage and salary workers. All self-employed workers are excluded, both those with incorporate businesses as well as those with unincorporated businesses. Updated population controls are introduced annually with the release of January data.

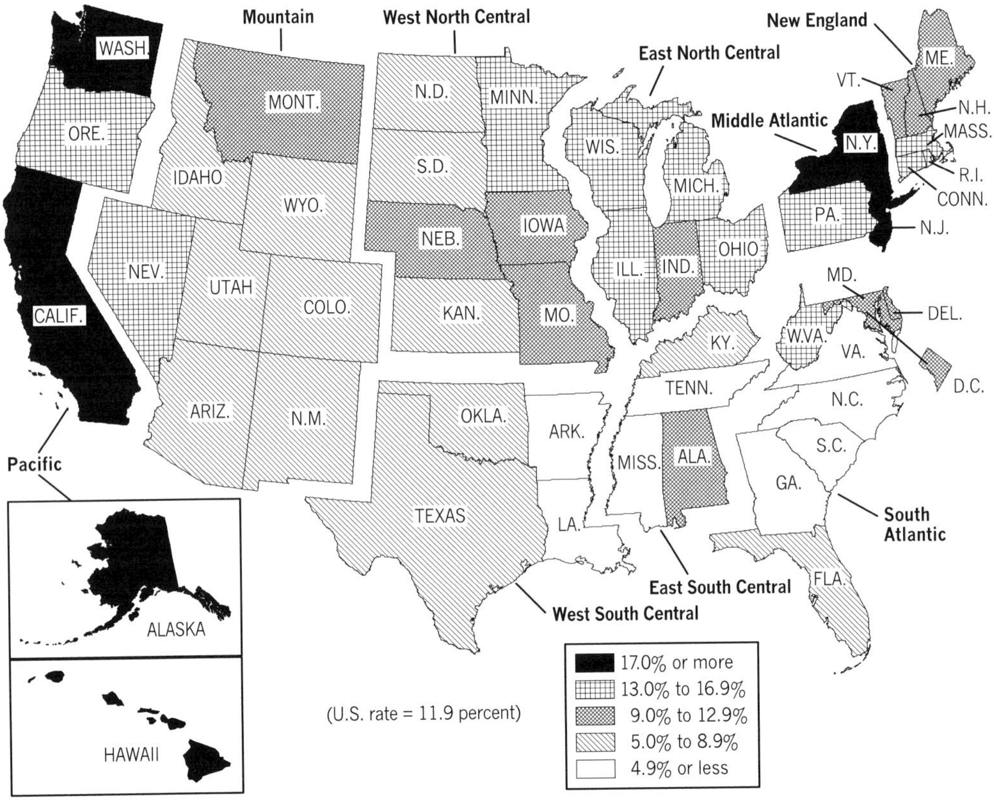

Figure 3–5. Union membership rates by state, 2010 annual averages

Arguably, the changing workforce is the single most important factor contributing to the decline in union membership. Often, students new to labor relations will claim that the work of the early labor movement—through better employment laws, better management skills from owners, and the lack of understanding of a union—is instead the result of the decline in membership. A common impression of students is that the union is inept, leaving the work for someone else to do. Workers who have held elections today to join unions have voiced concerns over topics such as wages not keeping up with inflation, poor working conditions (physically and emotionally), and inconsistencies within the company.

Granted, because many of the issues that workers faced in the early 1930s and had fought for are now written into law, workers today no longer need to fight for these same issues. Thanks to the hard work of the labor movement, many people today enjoy workplace protections. The goals of many unions today are to preserve their existing strength and to influence workplace laws and regulations. The labor movement serves as a watchdog to ensure that workplace laws are not amended such that less protection is given to the workers.

Today's labor unions have played an active role in advancing workers' rights. The Americans with Disabilities Act, the Equal Employment Opportunity Act, and the Occupational Safety and Health Act are widely recognized examples of this activism. Some workers feel that they have ample workplace laws and therefore no longer need unions, as evidenced by their declining membership. Conversely, others feel that if the labor movement gave up the fight for labor, then in time, management would reverse the direction of employees' rights. The union can serve as a means for communication between the workers and management. The process should not be a one-way street; rather, it should be a partnership.

Section 14(b) of the Taft-Hartley Act affirms that states can enact Right to Work laws. Right to Work laws provide protection to the individual employee to choose to belong to a labor union, or not to join or financially support a union; that is, the employer and the union cannot compel the individual employee to join a union as a condition of employment. In Right to Work states, employees are not required to join the union, yet they will receive the same level of benefits that the exclusive bargaining unit (the union) required in the collective bargaining agreement. When employees choose not to join a union and receive the same benefits from the union this is referred to as *free riding*.

The National Institute for Labor Relations Research states that between 2000 and 2010 the percentage of real growth in private sector employee compensation stood at 11% for Right to Work states compared to only 0.7% in forced-unionism states, with the national average standing at 4.3%. The cost of living–adjusted compensation per private-sector employees (2010) was listed at $56,575 for Right to Work states and $55,420 for forced-unionism states, and the national average compensation at $55,896.[14] More and more evidence is now showing that Right to Work laws are economically beneficial. David Littmann, senior economist for the Mackinac Center for Public Policy, presented the following testimony in front of the Michigan House Tax Committee on Restructuring: "Economic growth in right-to-work states has so convincingly and consistently eclipsed the average growth for non-right-to-work states that it makes the whole argument for workplace flexibility a non-controversial subject."[15] Those who support Right to Work laws will argue that the every person should have the right to work for a living, regardless of union participation. It is interesting to note that an employee can refuse to pay union dues because of a legitimate religious objection as supported by the NLRA, Section 19 and actions by the U. S. Supreme Court (*Machinists v. Street*, 367 U.S. 740 [1961]; *Abood v. Detroit Board of Education*, 431 U.S. 209 [1977]; *Ellis v. Railway Clerks*, 466 U.S. 435 [1984]).

Employees that are in non–Right to Work states also have a right to join or not join a labor union; however, in most cases, the employee who chooses to leave the union or chooses not to belong to the union will be required to pay the union a sum called a *user fee* or an *agency fee*. This fee is based on a percentage of what it costs the union to bargain the collective bargaining agreement that covers the individual employee. The burden is on the union to demonstrate what the cost is for bargaining. The union must establish and explain how they determined the fee. The individual employee reserves the right to challenge the fee. Special rights have been granted to individuals whose religious beliefs prevent them from joining or contributing money to a labor union.[16]

The following 22 states currently have Right to Work laws:

- Alabama
- Arizona
- Arkansas
- Florida
- Georgia
- Idaho
- Iowa
- Kansas
- Louisiana
- Mississippi
- Nebraska
- Nevada
- North Carolina
- North Dakota
- Oklahoma
- South Carolina
- South Dakota
- Tennessee
- Texas
- Utah
- Virginia
- Wyoming

Louisiana

TITLE 23. Labor and Workers' Compensation, Chapter 9: Miscellaneous Provisions Part V. Right To Work La. Rev. Stat. Ann. §§ 23:981 through 987

§ 983. Freedom of choice

No person shall be required, as a condition of employment, to become or remain a member of any labor organization, or to pay any dues, fees, assessments, or other charges of any kind to a labor organization. (Enacted July 9, 1976; effective October 6, 1976.)

Public partnership

President William Clinton issued EO 12871, which created the National Partnership Council (NPC). The NPC was made up of federal agencies and union members representing federal employees (7 management and 4 union members make up the 11-person council). The goal of this program was to craft a partnership between public sector labor and management, to improve how the government delivers its service and to foster better cooperation between labor and management.[17] The order states, "The involvement of Federal Government employees and their union representatives is essential to achieving the National Performance Review's Government reform objective."[18] This was the first time that government took such a bold step in fostering a partnership between labor and management. EO 12871 provided the guidance, monitoring, and evaluation of the partnership. Thus, it was left up to the individual labor-management teams to identify problems and develop workable solutions to provide better service to the people they served. As labor-management teams started to develop this partnership, issues over statutory changes of bargaining and union security were identified. Union security has been the lifeblood of the union, and any attempt to weaken this or lessen union security has been met with determined resistance from the labor unions nationally. Labor and management agreed to disagree on this issue, and they moved on to developing a partnership. In October 1996, the NPC issued a written report to President Clinton about the status of EO 12871:

> Since the Executive Order was issued, partnerships have continued to grow throughout the executive branch. There are well over one million bargaining unit employees in the federal government. Seventy percent of bargaining unit employees (up from 55% in 1994) are represented by partnership councils. Seventy-two percent (up from 53% in 1994 are covered by partnerships.[19]

Over the next few years, the general consensus was that the partnership was working and that the formal complaints between labor and management had decreased resulting in a better relationship. Weighing in on whether partnerships work, Holley, Jennings, and Wolters have made the following assertions: "first, partnerships can improve labor-management communications; second, [partnerships] can also contribute to the enhancement of the labor relations climate; third, partnerships has the potential to improve agency performance; fourth, management committeemen and commensurate union strength are essential to realizing the potential of collaboration; fifth, partnerships contribute to cost savings due to reduced workplace disputes; sixth, partnership were effective when the efforts were precisely targeted"[20]

Just prior to George W. Bush's taking office, the Heritage Foundation issued a report calling for the next U.S. president to end EO 12871. The report claimed that the goal of efficient government was not achieved and that the process only slowed the decision-making process for government. On February 17, 2001, President Bush signed EO 13203, calling for the dissolution of the NPC. President Bush's EO stated,

> Section 1. Executive Order 12871 of October 1, 1993, as amended by Executive Orders 12983 and 13156, which established the National Partnership Council and requires Federal agencies to form labor-management partnerships for management purposes, is revoked. Among other things, therefore, the National Partnership Council is immediately dissolved.[21]

In addition, EO 13203 revoked any and all orders, rules, regulations, guidelines, or policies relating to EO 12871, thereby dealing it a fatal blow. Mark Roth, general counsel for the American Federation of Government Employees, called it "a sad day for Americans because the results of labor-management partnerships are well-documented. They have resulted in savings, efficiency, higher productivity and a greater focus on customers."[22]

EO 12871 has been revised and implemented at the state level. On November 2, 2007, Governor Bill Ritter of Colorado signed an EO, Authoring Partnership Agreements with State Employees. The aim of the EO is to bring efficiency and innovation to government through a labor-management partnership. At this point, it is still too early to see the full effects of such a partnership. The prior labor relations between the two parties will have a direct and significant impact on how quickly results can be seen with a new partnership. Even without a state or federal mandate, this type of change could be implemented at the local level. This is not just a union thing; this is about employees and management reengineering and rethinking how they deliver a service to the community. For example, the local volunteer fire department could sit with the fire chief, town manager, and selectmen to set up a partnership program just for the fire department or this could be used townwide. It can be used for full-time employees, part-time employees, or volunteers. It can be for union and nonunion employees. What matters most, is that each side must be willing to buy into the partnership.

On a smaller scale, some fire departments have implemented such programs. They may be called *safety committees* or *labor-management committees*. In some cases, the two parties would mutually agree to meet and discuss problems that might arise from the employment relationship. What can become an issue with the committee is that some items presented to the committee can encroach on items that are required to be discussed at the bargaining table; care must be taken to ensure that the committee's action will not undermine the collective bargaining process, but rather should enhance the process. One way is to have the two parties consent to the following condition: "It is agreed and understood that policies and procedures related to terms or conditions of employment are appropriate matters for consultation between the Parties, providing however, that neither Party waives or relinquishes their right to negotiate mandatory subjects of collective bargaining providing, however, that the Parties may mutually agree to discuss any subject matter not otherwise included." This protects each side from straying to the area of collective bargaining. Generally, if you can get past this, then you are well on your way to developing a solid labor-management partnership. The aim of a partnership should be to foster open and honest communication about the policy and other employment-related matters. Discussing pending or potential issues before they become a matter of contention can be beneficial to both labor and management.

Since 1992, the IAFF and the International Association of Fire Chiefs have partnered in developing the Labor-Management Initiative Program. The program consists of two parts: first, a two-day training program for the fire chief and the union president; then, after six months, another two days of training (four days total). All of the training is conducted by a team comprising the fire chief and the union president.[23]

Labor-Management Initiative Program benefits include the following:[24]

- Improved labor-management relations
- Increased trust between the fire chief and the union president
- Improved interpersonal communications between labor and management
- Increased involvement and cooperation from all members of the organization
- Increased focus on goals benefiting both the organization and its members
- Reduction in adverse actions that affect labor-management relationships
- Improved service to the community

Just because a state does not have public sector collective bargaining laws does not mean that there is no need for a union. In the 12 states that do not have collective bargaining laws on the books, state firefighters associations/unions still represent firefighters. The City of Little Rock, Arkansas, signed a memorandum of understanding with Little Rock Firefighters Union, Local 34, despite the state's having no collective bargaining laws for the public sector. In fact, three-fourths of Little Rock's public-sector employees are covered by collective bargaining agreements—a process that has been honored for more than 20 years.[25]

The 110th session of Congress was considering the Public Employee-Employer Cooperation Act (H. Rep. 980) in an effort to have the federal government require that all states enter into collective bargaining agreements. Some have argued that Washington should not be making a one-size-fits-all bargaining law—instead, it should be left up to the individual states. Additionally, a law cannot mandate a labor-management relationship.[26] The 111th session of Congress is now considering a new version of the bill, the Public Safety Employee-Employee Cooperation Act of 2009 (Sen. 1611).

Labor-management relationships are necessarily built on trust, respect, and open and honest communication. A law may require two parties to form a partnership, but no law can require that the partnership work in harmony and produce results; this is often dependent on the relationship that labor and management had prior to the forming of the partnership. Many factors will come into play when forming a partnership, such as personalities, power, trust, communication and politics when it comes to a partnership or building a labor relations program. Look at your own organization: How willingly would labor and management enter into a partnership? How about if they are forced to do so by law?

Student Activity

Track the current status of the Public Safety Employer-Employee Cooperation Act. The first time this form of legislation was introduced was in the 105th Session of Congress. Report on current developments on the bill, revisions, or new legislation. In addition, research current bills pertaining to labor-management relations for public sector employees at the federal level or the state level.

Some labor-management relations have been adversarial at best. A few years ago, public administrators realized that more could be gained from the relationship with the respective unions if they entered into a partnership.[27] The hard part was selling this new idea to labor and management; after years and years of adversarial relationship, getting them to buy into such a program was not going to be an easy task.[28]

Making a law that state and local agencies must bargain collectively in no way guarantees that labor and management will get along. Take a look at the number of firefighter unions who have collective bargaining as a right under state law and you will find that some locals still have an adversarial relationship. In some cases, the relationship is so bad that the union takes a no-confidence vote in the fire chief.

In Saint Paul, Minnesota, collective bargaining has been utilized for many years with the firefighters and other city employee groups. A collective bargaining agreement did not prevent the relationship between the firefighters and the chief from becoming "contentious" and worsening to the extent that the city hired an outside firm, TriData, to conduct an audit of the Saint Paul Department of Fire and Safety Services. The audit results showed evidence that the relationship was in a critical stage: the fire chief was antagonizing the union; the union was filing grievances on everything it could; and union members were contacting political leaders in the city. The TriData report states that, "like many firefighter groups, Saint Paul firefighters are very active in the internal and external workings of the department. They have been very successful in the political arena, being major supporters of the Mayor and several City Council members."[29]

The Town of Yarmouth, Massachusetts, has maintained a collective bargaining agreement with the Yarmouth Professional Firefighters for many years. However, that did not prevent the relationship between the fire chief and the union from becoming adversarial. In an effort to tell the public and town officials, union members took a vote of no-confidence in the fire chief. The vote of no-confidence is a tactic that some unions have been using in an effort to discredit the person holding the position. Most votes of no-confidence have yielded little to no results. The state union or national union representative will often go public with a statement that the fire chief is the worst administrator to be hired in the fire service and will often use outrageous terms. In Yarmouth, nothing was done to the fire chief; if there was any result, it was that going public had fueled the already negative relationship between the town, the fire chief, and the union. In fact, within 24 hours of the union press release, town selectmen took a vote of confidence in the fire chief. Regardless whether a labor organization has a collective bargaining agreement, the labor-management relationship can become very negative. Months later and with the fire chief still working in the department, it is too early to determine what, if any, impact the vote of no confidence has had on the bargaining relationship. Building a positive relationship—or, better, a partnership—involves much more than the paper on which the collective bargaining agreement is written.

Strikes in the public sector

> Taking away the right to strike is a bit like eliminating the vapor safety valve on a boiler. Employees need to know that they have this means of relieving their frustrations and internal tension—even if they never use it. Otherwise an explosion is inevitable.[30]

Public and private labor unions in the late 1800s and the early 1900s did not have to deal with no-strike laws at the federal or the state level. It was not until 1935 that Congress passed the NLRA, giving private sector unions recognition to enter into collective bargaining. Prior to being prohibited by the Taft-Hartley Act of 1947, public sector strikes were occasionally used to resolve labor conflicts. In 1889, the Ithaca, New York, police went on strike as a result of wage issues. In January 1903, the Tampa, Florida, firefighters walked off the job as a result of personal abuse.[31] Regardless of the Taft-Hartley Act, public sector strikes will continue to occur.

Unions have used strikes as an economic weapon ever since their earlier days. In the public sector, it comes down to politics, not economics. Early on, the public sector was allowed to strike much like their counterparts in the private sector. Carrell and Heavrin have stated that "the right-to-strike issue is to the public sector what the right-to-work issue is to the private sector.[32] Many states have been reluctant to grant public sector unions the right to strike.

Without the power to strike, the public sector needed other means to gain support for their wants and needs at the collective bargaining table. Craig Olson, of State University of New York at Buffalo, supports the public sector strikes and claims that, "despite the high costs of strikes relative to the direct costs of hiring an arbitrator to resolve interest disputes, few parties in the private sector voluntarily agree to substitute interest arbitration for the right to strike, which indicates that the parties derive tremendous benefit from being able to determine their own future free from the unpredictable decisions of an arbitrator."[33] At present, 38 states have legislation that prevents public sector employees from striking, and 12 states have made it legal for some public sector employees to strike. For example, in 1967, the State of New York passed the Taylor Law, which mandates that public sector collective bargaining prevent strikes.

State of Florida

Chapter 447, Part II—PUBLIC EMPLOYEES

447.505. Strikes Prohibited

No public employee or employee organization may participate in a strike against a public employer by instigating or supporting, in any manner, a strike. Any violation of this section shall subject the violator to the penalties provided in this part.

Source: http://www.leg.state.fl.us/statutes/index.cfm?App_mode=Display_Statute&URL=Ch0447/ch0447.htm

State of New York

Public Employees' Fair Employment Act—Civil Service Law, Article 14

The Taylor Law

Grants public employees the right to organize and be represented by a union of their choice, or to refrain therefrom; requires public employers to negotiate with such unions concerning terms and conditions of employment of employees; establishes impasse procedures for the resolution of disputes in negotiations; defines and prohibits improper practices by unions and public employers; prohibits strikes.

Source: http://www.perb.state.ny.us/stat.asp

Many states have outlawed public sector strikes in an effort to maintain public service and safety. In turn, many of the states have elected to have labor and management turn to third parties to settle bargaining impasses such as binding interest arbitration, final offer arbitration, or alternative dispute resolution. The issue of strikes in the public sector has a long history as an adversarial relationship simmering and waiting to blow. Labor and management each tempt the opposite side to make the first move.

While I was a member of the Yarmouth (MA) Professional Firefighters Union, the town and the union could not agree on a collective bargaining agreement, and the union resorted to conducting *informational picketing*. Each week, the union gathered before the selectmen's meeting and walked up and down the sidewalk in front of town hall to display to the public our frustration of not having a signed contract. The union membership hung a billboard sign listing the number of days the firefighters had been without a contract on a boat trailer, parked in front of the town hall, as a reminder to the public and officials that the firefighters did not have a contract. At one point, during the Professional Firefighters of Massachusetts Annual Conference in Hyannis, Massachusetts, several hundred firefighters from across Massachusetts boarded buses to stand in solidarity with the Yarmouth Professional Firefighters Union at the selectmen's meeting. The pressure was on the political leaders of the community and the fire chief to end the stalemate. An article published by Anne Ross in 1969 quotes William Buck, president of the IAFF, discussing strikes in the public sector:

> Certain arbitrary public officials knowing that we cannot and will not strike because we voluntarily gave up the right in 1918 when we were founded, have certainly taken advantage of the professional firefighters across the land. As a matter of fact, the record will show we have been exploited by such arbitrary public officials who often times dared us to strike, knowing that we would not.[34]

Numerous firefighter organizations (union and nonunion) have taken their issues public in an attempt to gain public support. Public support translates into political power; political power translates into obtaining items of interest to the group. If collective bargaining falls to the political pressure at the current time, it may not be in the best economic interest of the community being served. If public safety employees take action to go on an illegal strike, they risk damaging the labor relations and possibly damaging the public image of the union.

Interestingly, Raytheon Missile Systems and 26 firefighters from the International Association of Machinists and Aerospace Workers Local 933 embarked on a 10-week strike that ended in January 2007. At issue were wages, retirement, and disability programs.[35] Privately owned fire departments do not fall under individual state labor laws. In this example, the local union and the employer (a private company) would come under the NLRA. In the private sector, the labor union can go on an *economic strike* that involves wages, benefits, and working conditions. The economic strike must be over mandatory bargaining items (wages, health insurance payments, pension contributions, work schedule, just-cause discipline provisions, grievance arbitration, and drug testing). The labor union may also choose to go on an *unfair-labor-practice strike*. In this case, the labor union refuses to work as a result of the employer's unfair labor practices. The economic strike and the unfair-labor-practice strike each constitute an action taken by the union. The private sector is granted protection to conduct strikes under Section 7 of the NLRA. Section 7 of the act states, in part, "employees shall have the right…to engage in other concerted activities for the purpose of collective bargaining or other mutual aid or protection."[36] Note that economic strikers can be permanently replaced. The next type of strike in the private sector is a *lockout*. In the case of a lockout, the employer is taking the action, notifying the employees not to come to work until an agreement is signed by the two parties. Regardless of the strike action that labor or management takes, they are bound by legal limits for the different types of strikes. For example, a union may choose to go out on a noneconomic strike, they will not be afforded any protection under the NLRA for such action (drug/alcohol screening for applicants, interest arbitration, bargaining unit expansion, benefits for retirees, or contract ratification).

Student Activity

Basic Guide to the NLRA

To obtain a better understanding of the framework pertaining to the NLRA, review the following document:

http://www.nlrb.gov/nlrb/shared_files/brochures/basicguide.pdf

You may print out a hard copy and compare and contrast your own state labor laws to the NLRA.

After examining legal and illegal strikes in the United States, Stern and Olson found that interest arbitration in place of strikes showed the fewest illegal strikes. Interestingly, states with collective bargaining legislation seemed to have the most illegal strikes, when the strike was replaced with mediation, fact-finding, and/or interest arbitration.[37]

What should be at the forefront is creating a better method of resolving bargaining impasses and labor issues, not just legislation banning strikes by governmental employees. Not one firefighter would agree to a strike if it would compromise the public safety, health, and/or welfare that each of us so proudly stands for. Only out of frustration will firefighters and other public employees or join informational picket lines. It is clear that the public sector is in need of a means to resolve these issues. As outlined in this chapter, not all states agree with allowing or disallowing public employees the right to strike. Regardless of this argument, labor organizations and the government need an effective method of dealing with the concerns on each side.

North Carolina

§ 95–98.1. Strikes by public employees prohibited

Strikes by public employees are hereby declared illegal and against the public policy of this State. No person holding a position either full- or part-time by appointment or employment with the State of North Carolina or in any county, city, town or other political subdivision of the State of North Carolina, or in any agency of any of them, shall willfully participate in a strike by public employees.

One author who believes that public sector employees should be allowed to strike is Robert Hebdon, of Cornell University. Hebdon has asserted that "collective bargaining without the right to strike may have serious unintended consequences. It may prevent the adjustment of expectations necessary for compromise; stifle valuable information on the preference of the parties; lead to major explosions of pent-up pressure; and be more difficult to regulate due to the lack of clearly defined objectives."[38] In the private sector, a union can make only so many demands of a company before the company decides to agree to the demands, move the company to a less unionized state, move the company out of the U.S. borders, or—the worst-case scenario—close its doors. In the public sector, the government does not go out of business, and relocating is not an option.

Appointed or elected government officials have been charged with providing the most cost-effective and expedient service possible. Political pressure may lead government agencies to give in to the union demands; however, giving in to all of the union demands will only hurt the long-term relationships. Before long, the demands will become excessive, which will be fought with excessive tactics, leading the public to develop a negative view of the demands and the tactics of the union. So, in a nutshell, unionization in the public sector involves political input, control, and support. An understanding of politics is a prerequisite to gain insight into the political powers within the public sector labor movement.

The inability of many emergency service unions to strike has necessitated other means of resolving negotiation impasses, such as *binding interest arbitration*. When labor and management cannot resolve their differences during negotiation, they often turn to a third party who is neutral to settle the impasse. Sometimes, the third party can bring labor and management together for better communication, and the two sides can settle the agreement without the neutral third party issuing an official ruling in the case. In other cases, the third party renders a final and binding decision resolving the impasse. The three most common methods used to settle impasses are as follows:

- Mediation
- Fact-finding
- Interest arbitration

These three items will be discussed in much greater detail in chapter 5. Ultimately, many public sector unions need another means to resolve negotiation impasses, rather than taking a strike vote. The closest thing that public unions have is informational picketing. Employees have a right to take their labor dispute public in an effort to gain public support over the manner in which their employer is treating them. The picketing must be peaceful and not involve threats, intimidation, and violence. Some unions have gained support by producing bumper stickers, pins, and signs to seek the public's support. In the event of an impasse, personnel management will often be affected. The decrease in morale and increase in sick time present challenges for supervisors and administrators in the fire service.

Labor Laws in the Public Sector

The issue of interpreting labor laws, regulations, and court cases has been a hot topic in the public sector and especially the fire service. As noted in chapter 1, the first fight that public sector employees faced was whether the government had to bargain with employees. Many states have individually addressed this issue by adopting legislation allowing collective bargaining agreements with public sector employee groups. Public safety employees are often allowed other means to settle contract breakdown compared to the private sector, which has the right to strike.

In the event that a labor disagreement cannot be settled between the two parties, the disagreement may be directed to the court system for relief. With court cases dealing with public sector labor laws, the courts must first determine, on a case-by-case manner, what rights and privileges are granted to the employee and the employer (government). The courts may look to other decisions/rulings for guidelines in such a legal case. In many cases, the courts only have to look back no more than 40 years for public labor law precedents. Some consider the field of labor law to be fluid—that is, it must be constantly watched for recent decisions that will affect labor relations. Personnel management in the fire service is governed by numerous legal limitations. The standard a couple years ago is no longer the standard today, based on legislative or legal standings. Followers of labor relations must stay current on legal rulings to ensure compliance with new rulings.

Originally, the government claimed that the *sovereignty doctrine*—the supreme, absolute, and uncontrollable power that a state has in governing—was the foundation for denying the rights of public employees to bargain collectively.[39] Therefore, the government is sovereign and it shall retain its management right—specifically, the right to hire and fire.

The U.S. Constitution has granted the power to Congress to craft laws that regulate interstate commerce and, in doing so, to regulate labor relations. These laws can protect the employer or the employee. A new law passed by Congress may be challenged in the court of law. It is up to the courts to determine if this law has legal standing, if any existing doctrine can be applied, and, finally, if the law applies to the case at hand.

The fire service is no different from any other business when it comes to labor law. Many decisions are bound by legal standings, and in some cases, decisions are dictated by law. For example, the use of a fire department computer system to send and receive e-mail has seen several legal challenges. Some states have viewed the e-mail communication written on a fire department computer to be a form of public record that can be made available under the Freedom of Information Act. This would mean that the press and other interested parties could have full access to any written communication prepared on a fire department computer and/or text message sent via a fire department phone.

It is recommended that fire departments establish a written social networking/media policy, rather than restricting firefighters' activities on social networking and media sites. The goal should be to educate the firefighter on the positive results of promoting the fire department. All employees should receive formal training on how to conduct themselves in a professional manner on such Web sites, which are often found on the department computers and phones, as well as on firefighters' personal cell phones. Care must be taken when photographing an incident; if firefighters later upload the pictures to a networking site, this could place you and your department in the middle of a legal debate: Have you violated an individual's right to privacy by taking his/her photograph? Can a law enforcement agency subpoena your pictures for a court case?

Personnel managers, regardless of the type of department they are from, need to be aware of the labor relations laws governing their department and employees. The majority of labor relations cases involve items such as wages, hours worked, workplace safety, employee benefits, unions, and discrimination. The U.S. Department of Labor enforces more than 180 labor laws and regulations that cover about 10 million employers and more than 125 million workers (fig. 3–6).[40] Many of the labor laws at the federal and state level are commonly applicable to the fire service, firefighter candidates seeking employment, firefighters currently employed, and retired firefighters. Many of the laws discussed in this book will provide you with a brief description and should not be considered legal guidance; for authoritative information on these laws, you should consult the statutes and regulations themselves, and before taking any action, it is highly recommended that you seek the advice of an attorney. Personnel laws will impact career, call, and volunteer departments.

Fig. 3–6. U.S. Department of Labor, Washington, D.C.

Labor-management relationships represent a continual process, not just a one-step fix-it-and-forget-it. The first step is for labor and management to agree on their responsibilities and legal rights. If a group of employees approached management with concerns, management may agree that the group represents the whole. They may work on such issues as safety, better working conditions, and pay. In nonunion organizations, such issues may be covered in an employee handbook or in the departments' standard operating guidelines. In a union environment, working conditions are generally covered under the collective bargaining agreement (wages, assignments, benefits, discipline, and dispute resolution [grievances]). Regardless of what you call it, once it is written down and clearly articulated to all the members involved (posted in the written format, posted online, and/or given to each employee in electronic format), each side knows what is expected. It has been my experience that many grievances boil down to miscommunication or lack of communication over the given topic. Each side will argue its position based on its interpretation of the section and the meaning behind it. Finally, the written agreement must be utilized all the time, not just when one side wants to use it; it must be utilized fairly and equally across the board.

The following three points comprise the foundation of many employee-management relations:

- Recognition of the labor group and management
- Development of a handbook or labor agreement
- Administration of the written agreement at all times

Agreements can be very simple or they can contain pages and pages of detailed sections for the employees and management.

In some cases, the first step is not accepted easily by the governing body, and the issue will be taken to court. Accomplishing the first step will often set the tone on how labor and management will interact. The first step can quickly become an adversarial relationship. Over the years, many municipal leaders and elected municipal officials have not known how to address the issue of recognizing the labor group or even how to collectively bargain with a labor group. Fire service administrators should understand the rights, privileges, and limitations granted to a group of employees conducting an organizing campaign to win representation. Many of the rights and limitations are grounded in state labor laws and unfair-labor-practice cases. These guidelines established by the state and court cases will apply to the employee and the employer.

Generally speaking, once a group of employees has decided that it would like to be represented by a union, it may be asked to sign *authorization cards*, and if the group represents more than 50% of the workers, the employer may choose to voluntarily recognize the group as a union under the state law (if it applies to that state). The other option would be for the employer to deny the request and force the union to file an *election petition* with the state labor board. The state board may choose to grant the election if more than 30% comply. If less than 30% sign, then the board may require the employee group to obtain additional authorization cards.

Once an election has been approved, if more than 50% approve the representation of the union, then the state will approve the election and the employer is required to bargain with the newly elected union. In some cases, the government may have the option to have the case reviewed by the courts. Again, this is not always the case, depending on how state labor laws apply to organizing campaigns.

Publicly, fear was struck when public employees could join public sector unions, derived from the bad press unions had received over the years from the labor movement. The public did not want public employees crippling local, state, or federal government by striking. The AFL-CIO saw the opportunity for significant growth by incorporating public sector unions with them. As more and more public employees chose to join the AFL-CIO, the AFL-CIO created the Public Employee Department to address the needs of public employees.

The AFL-CIO estimates that the United States has about 15.4 million unionized people, with 10 million of those belonging to a local that is part of the AFL-CIO. On the federal and District of Columbia front, the American Federation of Government Employees reports that they have a total membership of nearly 220,000 members; no other public union represents close to that number. The American Federation of Government Employees also has a Federal Firefighter Steering Committee, whose goal is to track, monitor, and analyze legislative and policy issues. An issue that has confronted the American Federation of Government Employees is the very low number of dues-paying members compared to the number of employees covered by the collective bargaining. Holley and Jennings have stated that "Overall, 7.4 million public-sector employees are union members who pay dues to be represented by unions, but nearly 800,000 are not union members and do not pay dues ."[41] Keep in mind, even non-dues-paying members are represented during negotiations, contract administration, and grievance arbitration. If the union does not adequately represent the member, the members can sue the union for failing to provide representation.

It often comes down to how each side will respect each other. What must be remembered is: how did we get to this point? What is the goal of the employees and what is the goal of management? The aim is to try and bring the two very different goals of two distinct sides more in line with each other. Compromise and understanding will be needed to achieve a mutual agreement.

Many articles have been written and much research has been conducted by sociologists, psychologists, economists, and political science on the best way to address issues facing labor and management relations. Each has its own impact on the relationship at any given time.

Many states have laws governing labor relations, in addition to the federal laws pertaining to unions. In the public sector, collective bargaining is governed by the individual states. It is recommended that you review your local state laws pertaining to labor relations. For example, Texas law allows firefighters to enter into a collective bargaining agreement, if the local community accepts the provision of the law.

Note that the NLRA does *not* cover public employees in federal, state, and local government. In addition, the NLRA does not include railroad and airline employees or agricultural workers. The 1978 Civil Service Reform Act, Title VII, created the Federal Labor Relations Authority to act as a mediator and to clarify issues. Another significant difference is the scope of bargaining for firefighters compare to the private sector. In many cases, public sector unions tend to settle contract impasses and grievances differently than the private sector. We discuss this further in chapters 5 and 6.

Public safety bargaining

Public safety labor relations have been placed in the spotlight since the terrorist attacks on September 11th, 2001. Many Americans view firefighters as heroes. Some airlines will upgrade firefighters in uniform and military personnel as a way to say thank you. Americans have a warm spot for the local firefighters who they often feel are not compensated properly for the level of risk that they take day in and day out.

The American people have been affected by the post–September 11th events called the War on Terror. The issue of terrorism has found its way to the bargaining table for public safety personnel. Many cities and town have been facing issues at the bargaining table involving the terms and conditions contained in the collective barging agreements about the War on Terror.

On August 15, 2002, in New York City, about 10 thousand firefighters and police officers gathered in Times Square, to protest the contract negotiations then being undertaken with the city. At issue was what the firefighters and police officers claimed to be the Bloomberg Administration's unwillingness to properly compensate them with pay increases. Taking part in the public protest were actor James Gandolfini and Senator Hillary Rodham Clinton. The negotiating process was brought to the street in an effort to gain public support. The support from the public and political figures will result in political pressure on the city.[42] By taking the bargaining issues to the street, the firefighters are looking for the citizen with a heart and with logic. Face it, you may have done it, and many others have done it, we play the emotional trump card when it comes to budget cuts. By laying off firefighters, closing a fire station, or not approving a budget item may result in your family being left without proper fire protection. People tend to vote on the basis of emotions—for example, out of fear of being left without service. In other words, they want to know that, when they call 911, the fire department will be at their house within minutes.

Pay Factors

Mehay and Gonzalez have found that lack of competition with local communities restricted local public sector unions for wage increases.43 Historically, public sector wages have been considered much lower than in the private sector. Today, state and local governmental employees are now receiving the same or even better pay and benefits compared to the private sector. Dennis Cauchon, in *USA Today*, observed that "state and local governmental workers now earn an average of $39.50 per hour in total compensation....Private workers earn an average of $26.09 an hour."[44] Years ago, the belief was that government would pay its employees a low wage and make up for it in fringe benefits. Many of the fringe benefits today are hidden in other line items where the general public will not find them. For example, a hidden benefit may be that employees can retire early at a guaranteed rate as high as 80% of their salary. Braden and Hyland found the cost of providing public employee benefits 56% higher than in the private sector.[45]

According to the U.S. Department of Labor, in 2009, union members received $908.00 per week in wages compared to nonunion workers, who earned $710.00 per week.[46] Figure 3–7 highlights the trend in union wages and nonunion wages from 1983 to 2007. In 2007, union women earned 33% more than nonunion women, and union men earned 24% more than nonunion men. Among Hispanics who worked full time, union members earned $249 more per week than nonunion workers; this was the largest difference between union and nonunion median usual weekly earnings for any race or ethnic group.[47]

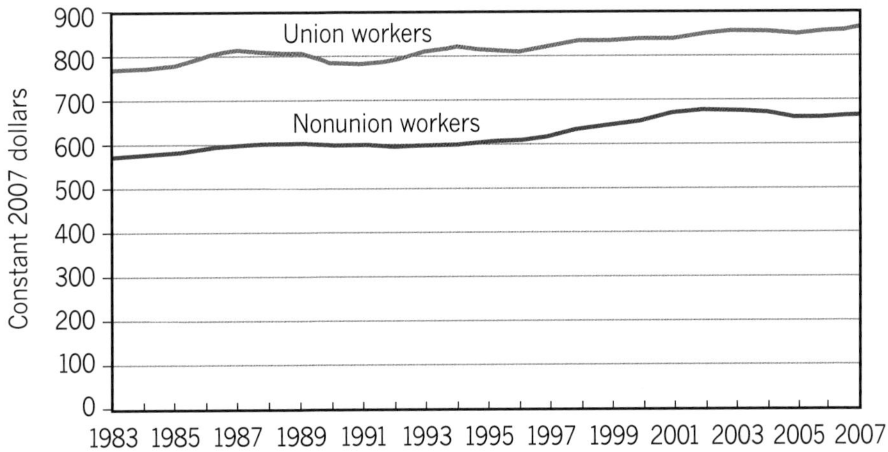

NOTE: The term "nonunion" refers to workers who are neither members of a union nor represented by a union on their job. Earnings are median usual weekly earnings of full-time wage and salary workers and have been converted to constant dollars using the Consumer Price Index research series (CPI-U-RS).

Fig. 3–7. Among full-time workers, union members had higher weekly earnings than nonunion workers, 1983–2007.

Wage theories in the public sector vary tremendously, and while some unions and communities believe in one collective theory, others may use a combination of two or three wage theories. The topic of wage theories is covered more in depth in other texts, and the reader is encouraged to research the topic to decide which theory works best.

Local media often only report the pay increase granted to public employees. Seldom do they get into the education incentives, additional vacation days, or sick-time buyback. Generally, the media will compare the firefighter increase to what the police or other public sector employees in the community received. Consequently, some communities will maintain a 3% wage increase across the board to all governmental employees, giving the appearance that all employees received the same compensation. However, when the true cost to each contract is broken down, you may find that the overall cost of the contract increased by 4.5% as a result of other fringe benefits. It is important to remember that the city/town officials are held accountable by the taxpayers in the community. Elected officials and appointed officials may want to give the general impression that they are being frugal with the budget and are offering only a small pay increase, preventing a tax increase. The bottom line is that community leaders must unofficially sell this contract increase to the taxpayers in the community; if taxpayers feel that the community leaders are just giving away large pay increases, then they may start revolting against the overall municipal budget.

Average Hourly Wages

Year	Public Sector	Private Sector
2009	$39.50	$26.09
2004	$34.72	$23.76

Source: U.S. Department of Labor, Bureau of Labor Statistics

Regardless of public or private sector, one wage theory is that the demand for labor is established by the demand for the product and/or service produced by that labor. If the demand for firefighter-paramedics is high, so must be the wages to attract firefighter-paramedics. Many fire unions have offered a separate article for specialty training/certifications. The contract may call for an additional 11% pay increase for members maintaining a paramedic certification. The pay increase will be based on the salary scale for the firefighters within the department. In other cases, departments will set a completely different salary scale for each specialized area. There is no single best way of establishing wage scales in a contract. It comes down to what works best for the community and the members. Other factors that come into play include local pay factors, housing, and living expenses.

Some argue that wages obtained by negotiations are the result of the strong negotiations tactics used by some unions. Some communities have a long history of paying low wages and accept the high employee turnover rate that results when people leave for better pay. Generally, if people are satisfied with their pay, benefits, and working conditions they are less likely to seek employment elsewhere. This holds true for union and nonunion employees. This may be linked to the institutional approach, taking into consideration how local politics can impact wages.[48] Regardless of the wage theory believed, one constant is the impact that political forces have on the outcome of labor relations.

Bargaining for wages can be one of the most challenging aspects the negotiator may deal with, aside from health insurance issues. In some cases, a community (town or city) may choose to bargain a multiyear contract and know what they need to budget in each of the next fiscal years. In other cases, communities are bargaining for past years and have been operating without a contract. This may be the result of the present economy and/or the current budget demands for the given community. Bargaining a multiyear contract can provide community leaders with solid numbers to budget next year, and when asked about the budget, the leaders can maintain that it is dictated by the union contract. One consideration in entering into a mulityear contract is the rate of inflation for the next three to four years.

Experts in the field of economics would argue that wages are established in the public sector based on three factors:

- An elected official's desire to be reelected or elected to higher office (fig. 3–8)
- Maintaining a stable tax base or lowering the tax rate
- Maintaining current services or increasing services[49]

Research by Belman, Heywood, and Lund has revealed that public sector union organizing may relate directly to increases in earnings.[50] This would suggest that unionized departments secure better wage increases than nonunionized departments.

Fig. 3–8. Local and state elections play a critical role in public sector labor relations. Organized labor representing public sector employees will aggressively engage in the political arena to protect collective bargaining laws.

In the past, parity pay with law enforcement was often an issue at the bargaining table. Even today, some fire unions are still seeking parity with law enforcement. Law enforcement and the fire service each view their jobs as highly dangerous, and for that we should be paid the same. Pay often comes down to the index of importance of a job. The more someone makes, the more important they are in our society.

Influence on Labor Relations

This chapter has outlined a number of issues that play a role in the labor relations process in the public sector:

- *Social.* The fire department budget is often approved by the taxpayers of the community. The fiscal stability of the community and the tax base will play a significant role in public sector budget.
- *Legal.* Ranging from the Civil Rights Act, Occupational Safety and Health Administration, U.S. Office of Personnel Management, Department of Labor, and Department of Justice to the individual state's laws pertaining to labor relations, the labor relations process is grounded in legal presidents and mutually agreed-on contracts. The manner in which a fire department recruits, hires, and promotes personnel must adhere to the current legal standard. The legal standards are not static and are constantly changing, causing the fire service administrator to constantly monitor the legal standing on a number of issues facing the department.
- *Local.* The community may have a direct impact on the labor relations process. The community may be for or against an issue currently being debated—for example, whether a fire department should be given a pay increase at a time when many other services in the community are being decreased because of budget cuts. If a local community has a very high unemployment rate, do you think the taxpayers will be receptive to giving a pay increase or to the firefighters complaints about not being paid for training?
- *Organizations.* Groups of organized employees such as labor unions can play a role in the labor relation process. In addition, nonunionized employees can also form an organization to deal with work-related issues with management. Taxpayer organizations can place significant pressure on elected and appointed municipal officials to support or defeat a labor relations issue. Municipal watchdog organizations can lobby for or against labor relation issues at the local level or the state level.

Final Thoughts

As you can see from this chapter, labor relations in the public sector is different than in the private sector. Many of the today's labor laws protecting public sector employees are the direct outgrowth of the private sector. The federal government has left it up to the individual states to decide how best they wish to regulate labor relations. So what you end up with is each individual state develops its own individual statues dealing with labor relations. The NLRA, as amended by the Taft-Hartley Act and the Landrum-Griffin Act, is considered to be the foundation for private sector labor relations in the United States.

Many public sector unions have been growing and will continue to grow, while the country sees a decline in private sector union membership. Public sector unions are here, and the trend in membership increases indicates that they will be around for a long time. It is in everyone's best interest to seek ways to improve the communication and working relations with the community and the union. One way that has been tested and supported is by way of forming a partnership. The partnership is not about any one group releasing any authority or losing control; rather, it is about sharing a common goal, providing the best service to our customers. Call it a collaboration, to provide the best possible service for less.

The U.S. Congress passed the Taft-Hartley Act, which prohibited public sector strikes, in 1947. Today individual states have taken steps to prevent public safety personnel from striking. Public safety positions are considered essential to government and therefore would put society at risk without them. Without the power to strike, the public sector needed other means to gain support for their wants and needs at the collective bargaining table. The union may choose to take a labor issue public in an attempt to gain public support and pressure elected/appointed officials into settling the disagreement.

Public sector unions today play a significant role in the political arena. They support candidates who they think will favor them in contract issues down the road. However, this political advantage can also be a disadvantage when the supported candidate is not elected into office. Local politicians can play a significant role in labor relations. The political climate in smaller communities can change overnight with an election.

Originally, the government claimed that the sovereignty doctrine allowed the denial of the rights of public employees to bargain collectively. The sovereignty doctrine is said to be the supreme, absolute, and uncontrollable power that a state has in governing. Many states have allowed individual communities to enter into negotiations with labor organizations. However, some states still do not allow a community to bargain with a labor organization.

Collective bargaining in the public sector is widely used in determining wages, working conditions, health insurance, management rights, and labor rights. With such a widely used process in the fire service, it would behoove all present administrators and aspiring administrators to better their labor relations skills. A theme emphasized in this book is that labor relations in the fire service applies equally to union and nonunion departments. How administration interacts with employees has a direct impact on the overall operation. Labor relations cannot be underestimated or neglected.

Key Terms

- Agency fee
- Authorization cards
- Binding interest arbitration
- Collective bargaining
- Economic strikes
- Executive Order (EO)
- Fact-finding
- Free rider
- Freedom of Information Act
- Informational picketing
- Interest arbitration
- Impasse
- Labor-management committees
- Labor-Management Initiative Program
- Labor Relations Process
- Lockout
- Meet and Confer
- Mediation
- Memorandum of understanding (MOU)
- Multilateralism
- National Industrial Recovery Act
- National Partnership Council (NPC)
- Negotiation
- Postal Reorganization Act
- Recognition
- Right-to-work
- Sovereignty doctrine
- Strikes
- Taft-Hartley Act
- Taylor Law
- Unfair-labor practice strike
- U.S. Constitution
- U.S. Department of Labor

Exploring Issues on the Web

American Federation of Government Employees: http://www.afgefirefighters.org/ (Of particular interest to fire service members is the Federal Firefighter Steering Committee, whose goal is to track, monitor, and analyze legislative and policy issues.)

American Rights at Work: http://AntiUnionNetwork.org/ (Web site launched by a national workers' rights organization on September 13, 2006, in response to the escalating public attacks on workers and their unions)

Chicago-Kent College of Law: http://www.kentlaw.edu/ilw/report.html (Quarterly report by Chicago-Kent College of Law, Illinois Institute of Technology, and the University of Illinois Institute for Labor and Industrial Relations)

Citizens Research Council of Michigan: http://crcmich.org/ (Providing factual, unbiased independent information on significant issues concerning state and local government organization and finance)

Civil Service Reform Act of 1978: http://www.eeoc.gov/abouteeoc/35th/thelaw/civil_service_reform-1978.html (Act reforming the civil service laws)

Cornell University: http://www.ilr.cornell.edu/mgmtprog/aboutUs.html (Offering workshops, certificates, and on-site training to meet the unique needs of professionals in the public sector)

Cyber Picket Line: http://www.cf.ac.uk/socsi/union/ (Directory of over 1,000 unions around the world)

Public-Private Infrastructure Advisory Facility: http://www.ppiaf.org/component/option,com_frontpage/Itemid,1/ (Providing grants to help governments explore public-private partnerships to expand and improve their infrastructure services and bring people out of poverty)

Texas Local Government Code, Chapter 174, Fire and Police Employee Relations: http://law.justia.com/texas/codes/2005/lg/005.00.000174.00.html (Texas law allows firefighter collective bargaining as long as the local community accepts the provision of the law.)

World Bank "Shrinking Smartly": http://www.worldbank.org/research/projects/downsize/ (Clearinghouse for researchers, development practitioners, and government officials concerned about the difficulties encountered in downsizing a large public sector)

References

1. Kearney, R. C. 1998. Labor law. In *International Encyclopedia of Public Policy and Administration*, ed. J. Shafritz. Boulder, Colo.: Westview Press. Pp. 1241–44.
2. Carrell, M., and C. Heavrin. 2007. *Labor Relations and Collective Bargaining*. 8th ed. Upper Saddle River, N.J.: Pearson Prentice Hall.
3. Holley, W., K. Jennings, and R. Wolters. 2005. *The Labor Relations Process*. 8th ed. Mason, Ohio: South-Western. P. 606.
4. Varone, J. C. 2008. Fire Officer's Legal Handbook. Clifton Park, N.Y.: Delmar Learning. P. 362.
5. U.S. General Accounting Office. 2002. Collective bargaining rights: Information on the number of workers with bargaining rights. http://www.gao.gov/new.items/d02835.pdf (accessed February 2, 2008). P. 6.
6. Ibid.
7. Varone 2008, p. 262.
8. Gershenfeld, W., and Cutcher-Gershenfeld, J. 1994. The scope of collective bargaining. In Handbook of Public Sector Labor Relations. New York: Marcel Dekker. Pp. 51–53.
9. U.S. Department of Labor, Bureau of Labor Statistics. 2008. Union members summary. Bureau of Labor Statistics. http://www.bls.gov/news.release/union2.nr0.htm (accessed March 17, 2008)
10. U.S. Department of Labor, Bureau of Labor Statistics. 2010. Economic news release. Bureau of Labor Statistics. http://www.bls.gov/news.release/union2.t03.htm (accessed June 3, 2010).

11. Ibid.

12. Ibid.

13. U.S. Department of Labor, Bureau of Labor Statistics. 2010. Economic news release. Bureau of Labor Statistics. http://www.bls.gov/news.release/union2.t03.htm (accessed June 3, 2010).

14. National Institute for Labor Relations Research (2011). Right to Work States Benefit From Faster Growth, Higher Real Purchasing Power—2011 Update. http://www.nilrr.org/node/233 (accessed November 22, 2011).

15. National Institute for Labor Relations Research (nd). Freedom to Associate of Necessity Means as Well Freedom Not to Associate. http://www.nilrr.org/node/67 (accessed November 22, 2011).

16. Herman, E. E. 1998. Collective Bargaining and Labor Relations. 4th ed. Upper Saddle River, N.J.: Prentice Hall. Pp. 65–66.

17. Holley, W., K. Jennings, and R. Wolters. 2009. The Labor Relations Process. 9th ed. Mason, Ohio: South-Western. P. 578.

18. Clinton, W. J. Executive Order 12871.

19. National Partnership Council. 1996. A New Vision for Labor-Management Relations: A Report to the President on Progress in Labor Management Partnerships. Washington, D.C.: Office of Personnel Management. P. 1.

20. Holley, W., K. Jennings, and R. Wolters. 2009. The Labor Relations Process. 9th ed. Mason, Ohio: South-Western. P. 578-579

21. Bush, G. W. 2001. Executive Order 13203. Revocation of executive order and presidential memorandum concerning labor-management partnerships. http://www.whitehouse.gov/news/releases/2001/02/20010221-1.html (accessed March 18, 2008).

22. Lunney, Kellie. 2001. Bush dissolves labor-management partnership council. Government Executive. http://www.govexec.com/dailyfed/0201/021901m1.htm (accessed February 1, 2008).

23. International Association of Fire Chiefs. n.d. Labor-Management Initiative Program. http://www.iafc.org/associations/4685/files/LMI_LMIinfoSheet.pdf (accessed June 10, 2010).

24. Ibid.

25. US-GAO 2002, p. 6.

26. Ibid.

27. Delaney, J. T. 1996. Work cooperation: Current problems, new approaches. *Journal of Labor Research*. 17 (1): 45–61.

28. Kearney, R. C. 1996. Managing relations with organized employees. In *Handbook of Public Administration*, ed. J. Perry. San Francisco: Jossey-Bass. Pp. 460–74.

29. TriData. 2007. Comprehensive management study of the Saint Paul Department of Fire and Safety Services final report. http://www.stpaul.gov/fireaudit (accessed March 19, 2008).

30. Cooke, S. T. 1983. A management view—the Canadian postal system. In IRRA 35th Annual Proceedings. Madison, Wisc.: IRRA. Pp. 94–101.

31. Cimini, M. H. 1998. 1982–97 State and Local Government Work Stoppages and Their Legal Background. *Compensation and Working Conditions*. Fall: Pp. 32–34.

32. Carrell and Heavrin 2007, p. 255.

33. Olson, C. 1982. Advances to impasse resolution: The use of the legal right to strike in the public sector. *Labor Law Journal*. 33 (8): 494–501. P. 501.

34. Ross, Anne(1969) Public employee unions and the right to strike. *Monthly Labor Review*. (March 1969) 92(3) Page 14-18.

35. Pallack, B. 2008. Raytheon firefighters authorize possible strike. *McClatchy-Tribune Business* News. February 3.

36. National Labor Relations Act, Section 7.

37. Stern, J. L., and Craig Olson. 1982. The propensity to strike of local government employees. *Journal of Collective Negotiations in the Public Sector*. 11 (3): 201–14.

38. Hebdon, R. 1998. Behavioral determinants of public sector illegal strikes. *Relations Industrielles*. 53 (4): 667–690.

39. Witney, F., and B. Taylor. 1996. *Labor Relations Law*. 7th ed. Upper Saddle River, N.J.: Prentice Hall. P. 437.

40. U.S. Department of Labor. 2008. Summary of the major laws of the Department of Labor. U.S. Department of Labor Office of the Secretary. http://www.dol.gov/opa/aboutdol/lawsprog.htm (accessed March 2, 2008).

41. Holley, W., K. Jennings, and R. Wolters. 2009. The Labor Relations Process. 9th ed. Mason, Ohio: South-Western. P. 571.

42. Rashbaum, W. 2002. Firefighters and police hold protest for raises. *New York Times*. http://query.nytimes.com/gst/fullpage.html?res=9C0DEFDE143DF935A2575BC0A9649C8B63&sec=&spon=&pagewanted=all (accessed May 6, 2008).

43. Mehay, S., and R. Gonzalez. 1986. The relative effect of unionization and interjurisdictional competition on municipal wages. *Journal of Labor Research*. 7 (1): 79–93.

44. Cauchon, D. 2008. State, local governmental workers see pay gains. *USA Today*. http://www.usatoday.com/news/nation/2008-02-01-civil-servants_N.htm (accessed November 18, 2011) February 2008.

45. Braden, B., and S. Hyland. 1993. Cost of employee compensation in public and private sector. *Monthly Labor Review*. May: 14–21.

46. Department of Labor 2010.

47. Walker, J. A. 2008. Union members in 2007: A visual essay. *Labor Review*. October: 28–38.

48. Reder, M. 1975. The theory of employment and wages in the public sector. In *Labor in the Public and Nonprofit Sector*, ed. D. S. Hammermesh. Princeton, N.J.: Princeton University Press. Pp. 1–48.

49. Kearney, R. C. 1996. Monetary impacts of collective bargaining. In *Handbook of Public Administration*, ed. J. Perry. San Francisco: Jossey-Bass. P. 79.

50. Belman, D., J. Heywood, and J. Lund. 1997. Public sector earnings and extent of unionization. *Industrial and Labor Relations Review*. 50 (4): 610–28.

LABOR RELATIONS AND COLLECTIVE BARGAINING

4

Learning Objectives

After reading this chapter, you should be able to:

- List the most common contract components of a fire service collective bargaining agreement.
- Recognize how bargaining units are determined in a union environment.
- Explain the importance of unfair labor practices in the public sector and the private sector.
- Recall the requirements for mandatory, permissive, and prohibited bargaining items.
- Discuss the elements of bargaining power as it pertains to collective bargaining in the fire service.

Collective Bargaining and Contracts

State law pertaining to labor relations will define the limits of collective bargaining for firefighters. State statute or local ordinance may allow collective bargaining, in which case the law/ordinance will govern the relationship between the labor union and the employer. State law may bring the two parties together, but it does not require them to agree on a collective bargaining agreement. The labor relations process involves the union (exclusive bargaining representative for the employees) and management (representing the employer) (fig. 4–1). Labor and management will jointly agree on specific terms of employment and how they will be administered. The agreed-on terms of employment will be written down and organized in a collective bargaining agreement. The collective bargaining agreement is a *contract—that* is, a legal agreement between management and the union for a specific time frame. Collective bargaining is a continuous process that requires labor and management to meet and confer on the terms of employment. The contract should be viewed as a living document, referred to on a daily basis and updated regularly. Collective bargaining agreements can be found in the public and private sectors across the United States and other countries. Collective bargaining has been refined over the years to a process that can result in mutual gains for labor and management.

Fig. 4–1. "The American Twins," a cartoon drawn by Chad Cullen depicting the relationship between labor and capital. The caption reads, "United we stand. Divided we fall."

Regrettably, the media frequently portray the bargaining process in a negative light when covering labor issues. For example, stalled and adversarial labor relations between union and management receive disproportionate attention in the press; in fact, the AFL-CIO reports that 98% of collective bargaining agreements have ended successfully.[1]

The terms of employment generally cover the following areas, pertaining to the union workforce:

- Wages
- Benefits

- Work rules
- Hours of work
- Discipline
- Job classification
- Seniority
- Promotion process
- Layoffs
- Work practices
- Grievance process
- Contract administration

Importantly, not all collective bargaining agreements will include the above items. Some states will limit what can be subject to bargained. Other contracts will be detailed and cover many more items not listed above.

Bargaining contract components

In this section, we will explore the items most often found in fire service collective bargaining agreements. Examples have been included to demonstrate the intent and the common verbiage. Several topics that may be included in a fire service collective bargaining are presented in table 4–1; however, not all contracts will contain these items.

Generally, collective bargaining agreements will start out with a *preamble*, stating who the agreement is between and the purpose of the agreement. In some cases, the preamble, whether short or detailed, lists the definitions used within the contract; the preamble is generally the first article or section contained in the collective bargaining agreement. In other cases, the preamble is listed first, and immediately following is the first article in the contract, spelling out the definitions used within the contract; once the preamble is written into the contract, the article stays the same from contract to contract. The following is a sample:

Preamble

This Agreement is entered into by and between the __(City/Town)__ hereafter referred to as the "City" or "Employer" and the __(Name of the Local Union)__ hereafter referred to as the "Union."

The general purpose of this Agreement is to establish wages, hours, and terms and conditions of employment for bargaining unit members and to provide an orderly procedure for the resolution of grievances.

Table 4–1. Bargaining contract components

Preamble	Contracts often begin with a Preamble stating the purpose of the contract and a section spelling out definitions used within the contract.
The Recognition Clause	This is an item usually appearing at or near the beginning of a labor contract. Its function is to give recognition to the union, as the representatives of the employees, entitled to bargain in their behalf.
Union Security	Once again, this is often one of the first topics in a labor agreement. While "negotiated," the union security clause essentially provides that the union continues to exist and will be allowed to fulfill its functions. It will also typically identify the type of union activity within the organization (i.e., closed or open shop).
Dues Check off Provision	Perhaps contained in one of the previous elements, many contracts do contain provision for "dues check off." This means, that with the individual employee's permission, the payroll office will automatically deduct membership dues from the employee's paycheck.
Union Rights and Responsibilities	Sometimes the above items are included in a broader article, such as Union Rights and Responsibilities. An article with this as the subject would also likely include discussion of union representatives (e.g. stewards and participants in labor-management meetings, time off from regular duties for union activities and other related items.
Employee Rights and Responsibilities	This article will address the basic rights of the individual as specified in state and federal law and the right to reasonable notice and opportunity to meet with the administration and confer on matters within the scope of representation.
Non-Discrimination/ Sexual Harassment/ Coercion and Intimidation Article	Many contracts today address these topics up front and in a stand-alone article. While typically reflective of policies already in place, these issues may be included within the contract particularly to address such issues as being the subject of or participating in investigations and absence from repercussion from reporting or testifying to such behaviors during a filing or during the investigation.
Other Organizational Policies	Many contracts will cite other policies of the organization by which union members agree to abide whether by name, reference, or summary.
Management Rights	Most labor contracts have an article dealing with management rights. How detailed or elaborate this article might be depends. Generally, management rights entail issues surrounding three main areas: the right to determine the mission, goals and objectives of the organization (i.e., direction of the organization); the right to determine the uses to which the material resources of the organization are put (scheduling employees, assigning employees, supervising employees, determining location of physical resources, etc); and the power to discipline for cause.
Compensation and Benefits	While anything related to "conditions of employment" is typically negotiable, compensation and benefits is typically one of the "biggest" items negotiated. It is usually one of the larger articles in the contract. There are a variety of issues falling into this domain. Some typical examples of items covered include wage and salary scales or schedules for the various positions represented, hours of work and work schedules, overtime issues (including such things as how overtime is determined), holidays, vacation days, sick leave, jury pay, other types of leave and perhaps, layoff or severance pay. (Often times the citing of these issues parallels those of the larger entity, assuming the service falls under a city or county jurisdiction).

Union/Management Meetings	Because labor management committees or groups have been so effective in resolving interim frustrations regarding the contract, contract interpretation, etc. many contracts now contain specific articles addressing union/ management teams, committees, and/or meetings. Such articles may specify which positions will serve as representatives for each side or how those representatives will be chosen, the frequency of such meetings, authority of such teams, and other issues relating to their organization and functioning.
Grievance Procedures	Virtually all negotiated contracts contain fairly elaborate protocol regarding discussion of grievances. Such articles usually start with a definition of a grievance, bases for grievances, and the employee's grievance rights/ union representation along with the actual steps / procedures in the grievance process. Many grievance articles also address such things as grievance files, the use or non-use of recording devices, and withdrawal of grievances. Sometimes contracts will specifically address the grievance rights – or lack thereof – for probationary employees.
Other Common Topics of Contract Articles	While not necessarily in each agency's contract, some other topics likely to appear include such things as educational and training benefits, health and safety issues, uniform allowances, use of jurisdictional vehicles, and a variety of other items. Sometimes these and miscellaneous items are contained within other articles.
Term of Agreement	Contracts typically conclude with stipulations regarding the availability of contracts (how they can be accessed) and the term covered by this agreement. Effective dates and ending dates are specified. While not universal, the duration of a typical contract is three years. Some cover one year; others remain in effect for five years. This is of course is followed by the signatures of union representatives and management representatives.
Appendices	There may be any number of attachments to the contract for reference.
Agreement and Ratification	Before the contract becomes final, it must be approved by the union membership and by management. Regarding union ratification, a ratification election is held (i.e., vote on the proposed contract) and generally speaking a majority of those voting in the election must vote in favor of the document prior to its official approval by the union. Similarly, "management" must approve the contract. In the public sector, of course, this is typically a city council, a board of commissioners, a county board, or other public representative. Once it is ratified and accepted by both sides, it becomes a legally binding document for labor relations for both sides during the designated period.

Source: FESHE Course: Personnel Management for the Fire and Emergency Services, Version 1.0, Winter 2007©

Generally, the next section in the contract is the *recognition clause*, which recognizes the union as the exclusive bargaining agent for the employees identified. In smaller unions, the unit may include firefighters up to and including deputy fire chiefs; in larger unions, the recognition clause may only cover line firefighters. Once the recognition clause is established, it does not change from contract to contract. Whether fire officers should be included in the bargaining unit has been a matter of long-standing debate, because of their supervisory/management role. State law may either require separate bargaining units for supervisors or prevent supervisors from collective bargaining; even if the supervisors unit and the rank-and-file unit are considered as separate bargaining units, they may still belong to an international union or its affiliate.

In the labor relations field, little weight is given to position title; rather, the deciding factor is the actual job duties. Face it: not all people holding the rank of captain perform the same job duties across the United States. Similarly, supervisors and managers perform diverse job duties that lack consistency. For example, the State of Montana Collective Bargaining Law defines a supervisor as a *supervisory employee*, meaning an individual who has "the authority on a regular, recurring basis while acting in the interest of the employer to hire, transfer, suspend, lay off, recall, promote, discharge, assign, reward, or discipline other employees or to effectively recommend the above actions if, in connection with the foregoing, the exercise of the authority is not of a merely routine or clerical nature but requires the use of independent judgment."[2]

The following is an example of a recognition clause:

Recognition Clause

The City hereby recognizes the Union as the sole and exclusive representative of all permanent full-time uniformed Fire Captains, and Firefighters, excluding the Chief of the Fire Department, hereinafter called the Chief, and excluding all other town employees, for the purpose of collective bargaining in respect to rates of pay, hours of employment and other conditions of employment. The term "employee" or "employees" as used hereafter refers to persons for whom the Union is the recognized exclusive representative.

In the public sector, many contracts will have a section called *union security* that does not require an employee to join the union but may require the employee to pay an agency fee or user fee that is less than the full membership dues to remain employed. In other words, the law allows the individual person the right to choose whether to belong or not to belong to a labor union. However, the person who chooses not to belong to the union must still pay a fee to the union for bargaining and grievance costs associated with the collective bargaining agreement. The fee established by the union can only used for the cost of negotiating and administering the collective bargaining agreement.

In some cases, the contract may include in the appendix the agency fee or user fee, so this section may need adjusting from contract to contract. Nonmembers of the union may ask to see how the fee was established by the union. A nonmember may challenge the fee, and the union may choose to lower the fee. In the event that a nonmember feels that the fee is too high and the union will not adjust the fee, that nonmember may file a dispute for final and binding arbitration with the contract or may file a complaint with the state labor board asserting unfair labor practice. The state labor laws will often be very clear on how agency fees are to be determined; you will note that the list of "approved union activities" is what can be used to determine the fee. The list of nonapproved activities cannot be used in determining the fee.

On a very limited bases, state law and local ordinances cannot prevent a employee from joining a employee organization or participating in activities for a employee organization. In the United States, the right of an employee to join an employee organization is grounded in the First Amendment to the Constitution. Consider this: The employee has a legal right to join an employee organization, but in some cases, the law does not require the employer to bargain with the employee organization (unless the state law requires the employer to bargain with the employee organization). When the law is silent on the issue of mandatory bargaining, the employer may chose to bargain with the employee organization.

Approved Union Activities

- Cost of contract negotiations
- Cost of filing and processing grievances
- Costs associated with union business meetings
- Cost of union publications
- Cost of litigation
- Cost of social events and member benefits offered to all members
- Cost associated with attending national/state conventions

Nonapproved Union Activities

- Cost of legislative lobbying
- Cost of benefits not available to nonmembers
- Cost of charitable contributions
- Cost of political donations

A sample of a union security article follows:

Union Security

Each Employee, who on the effective date of this Agreement, is a member of the Union shall, as a condition of employment, maintain his membership in the Union. Each present Employee or an Employee hired on or after the execution of the Agreement, as a condition of employment, at his option, shall become a member of the Union upon completion of the probationary period of the Fire Department, and maintain his membership in the Union, or in lieu thereof pay a service charge in an amount equal to the regular monthly dues of a Union member. Employees who fail to comply with this requirement shall be discharged by the Employer within thirty (30) days after receipt of written notice to the Employer from the Union.

Generally following the union security article is the *dues checkoff* article. The dues checkoff section spells out how the union dues and the user fee will be collected. The employee will sign a dues payment form to allow the union to automatically withdraw the union dues or user fee from the employee's paycheck and transfer it to the union account. Importantly, the employee must sign a release before the dues can be automatically withdrawn. From a union administration standpoint, if members did not have a dues checkoff option, the collection of dues and user fees would fall in the hands of the union administration to collect and track. The dues checkoff option streamlines the process and ensures a continuity of dues collection for the union. Generally the dues checkoff article will not need to be changed from contract to contract. Since dues checkoff is a benefit for the union, management will often agree to it in turn for greater flexibility in other areas.[3]

A sample of possible dues checkoff follows:

Dues Checkoff

The "City" agrees to deduct, once each month, dues and assessments in an amount certified to be current by the Treasurer of the "Union" from the pay of those employees who individually request in writing that such deductions be made. The "City" further agrees to withhold from employee's pay, special assessments with proper authorization from the employee, and remit the same to the "Union" Treasurer.

The goal of unions has always been to limit or clearly define management's decisions and power over employees. Limiting the power of management in the employee-employer relations plays a significant role in the labor relations process. The union will often pursue this goal of limiting management power if it will benefit the union or the employees. Management, by contrast, has long asserted that the operation of the fire department should rest in the hands of the fire chief and the municipality. After all, the municipality is paying a professional administrator to administer and efficiently deliver fire and emergency medical services to the community, so why would any municipality want to limit this ability? Over years of contract negotiations, in a give-and-take approach, management rights have been expanded for greater control of the department or limiting the management rights. When the municipality wants something from the union, it may offer to release certain management rights in order to regain specific benefits.

Arguably, just having a collective bargaining agreement limits the rights of the fire chief to run the department as he/she sees fit. Day-to-day operations and managerial decision-making in the fire department will often be rooted in the confines and limitations of the collective bargaining agreement. This can be viewed as a two-way street: the contract can place limits and controls on both management and labor.

Management rights articles can be found in private and public sector collective bargaining agreements. So, on a larger scale, one has to wonder: How much control does labor have over an industry? What role and extent should labor play in defining management rights? These questions remain to be answered. It has been estimated that 80% of all collective bargaining agreements have some section pertaining to management rights (also called "management security" or "management prerogative"); these statements may be extensive, or they may be short and limited. Conversely, 20% of collective bargaining agreements have no management rights sections, because management will not release, modify, or eliminate any of their rights in the contract.[4] Management rights play a major role in personnel management and in how fire officers assign work to firefighters.

Decisions to reduce or limit management rights can frequently be linked to politics. The goals and objectives of any service provided by the government are politically determined by the legislative body. The operations and management of these goals for service are rooted in laws and regulations. Since we, the people, elect the legislative body, it is noteworthy that the government can bargain away management rights for the delivery of service to obtain the goals and objectives of the service. So no matter what decision is made at the bargaining table, containing management rights may have a strong political presence. The people who are bargaining on behalf of the government are accountable to the voters. According to Joan Pynes, "public sector employers tend to have more discretion in exercising their management rights than private sector employers."[5]

Many in the field of public sector labor relations would contend that it is difficult at best to align a union's reasonability, the employee's desires, and the government's reasonability to provide a service to the community. Today, many labor unions and management are working together to find a common ground on which to maintain a positive relationship among all involved. This must begin long before the bargaining process.

The following is a sample of a possible management rights article:

Management Rights

It is understood and agreed that the Town possesses the sole right to operate the Fire Department and that all management rights are expressly reserved to the Town, but that such rights must be exercised consistent with the provisions of this Agreement. These rights include, but are not limited to, the following: the hiring, the assignment or transfer of employees; determination of the mission and objectives of the Fire Department; determination of the methods, means, and number of personnel needed to carry out the Fire Department's mission and objectives; introduction of new or improved methods, facilities, and/or equipment; and the scheduling of operation and shifts.

The above contract references have been provided to demonstrate sample contract language found in many labor relations contracts for fire departments. This is by no means required in contracts; rather, they are the most often found articles in contracts. Beyond these basic articles, contracts will vary significantly, so we will stop at this point with providing sample sections of contracts. As part of the next exercise, students are encouraged to review current labor contracts to assess the differences in the contracts.

Student Activity

Search the Web for fire department labor contracts. Compare and contrast two random fire department labor contracts. How do the management rights compare? Is one contract stronger than the other? What would you recommend changing if you sat on the union side of the table for management rights? What would you recommend changing if you sat on the management side of the table? Be sure to support your answers to each question.

Bargaining unit eligibility

The union will elect or appoint a group of members to a *bargaining committee* that will represent the union at the bargaining table. The bargaining committee will often appoint a *chief negotiator* to head the committee. Management will also form a bargaining committee and appoint a chief negotiator.

Under the law, the union has been granted the authority to represent all employees of the bargaining unit. This means that the union represents all employees, regardless of whether they belong to the union. It also prevents the union from attempting to cover non-unit employees. The bargaining unit eligibility will often spell out in detail which employees are covered by the contract (bargaining unit). This prevents management from pitting one union against another.[6] In smaller fire departments, all full-time firefighters may be covered under one bargaining unit, despite their rank within the fire department structure. In larger fire departments, there may be separate bargaining units based on the rank or level of supervision.

Most contracts will have a recognition article that specifies the union and who is considered under the unit. The following are examples of two such articles:

The City of Austin (Tex.) and the Austin Firefighters Association, Local 975

Article 3—Recognition of Association

The City recognizes the Association as the sole and exclusive agent for all firefighters pursuant to Local Government Code § 174.101.

The City of Seattle (Wash.) and the International Association of Firefighters, Local 27

Article 2—Recognition and Bargaining Unit

The City hereby recognizes the Union as the exclusive bargaining representative of all uniformed employees of the Seattle Fire Department, holding the rank or position of Recruit Fire Fighter, Fire Fighter, Fire Lieutenant, Fire Captain, Fire Boat Pilot, and Fire Boat Engineer.

When a collective bargaining agreement exists, not all employees in the fire department are allowed to belong. Generally, the distinction between *supervisory employee* and *confidential employee* is called into question during the recognition stage. The classification of an employee can determine whether that person is recognized by the collective bargaining agreement. Some states have defined each of these areas to better determine who is covered and who is not. For example, according to the State of Ohio Public Employees' Collective Bargaining Law (4117.01), a confidential employee is any employee who works in the personnel offices of a public employer and deals with information to be used by the public employer in collective bargaining or any employee who works in a close continuing relationship with public officers or representatives directly participating in collective bargaining on behalf of the employer. By contrast, in the State of Minnesota, Chapter 179A.03(17) defines a supervisory employee as a person who has the authority to undertake a majority of the following supervisory functions in the interests of the employer: hiring, transfer, suspension, promotion, discharge, assignment, reward, or discipline of other employees, direction of the work of other employees, or adjustment of other employees' grievances on behalf of the employer. To be included as a supervisory function, the employee's authority is not merely routine or clerical in nature, but must require the use of independent judgment. An employee, other than an essential employee, who has authority to effectively recommend a supervisory function, is deemed to have authority to undertake that supervisory function for the purposes of this subdivision. The administrative head of a municipality, municipal utility, or police or fire department and the administrative head's assistant are always considered supervisory employees. The removal of employees by the employer from a nonsupervisory appropriate unit in order to designate them as supervisory employees will require either the prior written agreement of the exclusive representative and the written approval of the commissioner or a separate determination by the commissioner before the redesignation is effective.

As a result of the definition of confidential employee and supervisor, the first order of business in labor-management relations is to identify exactly who is covered under the bargaining unit—which is why the recognition clause is often at the very beginning of a collective bargaining agreement. Naturally, labor wants the greatest number of employees covered by the collective bargaining agreement, while management would like to limit the number of employees covered.

Because the definition of confidential employee and supervisor is determined by each state, one state may allow assistant fire chiefs as well as line firefighters in a bargaining unit, while another state may break down the level of supervision into different bargaining units. Volunteer and reserve firefighters are not generally included in collective bargaining agreements with full-time firefighters. However, in some states, volunteer and call firefighters are allowed to enter into collective bargaining agreements, based on a state-to-state ruling.[7]

Scope of Collective Bargaining

The scope of collective bargaining in the public sector has a direct connection and is the outgrowth from the private sector. The Wagner Act (National Labor Relations Act) as amended by the Taft-Hartley Act (Labor Management Relations Act), § 8(d), defines the scope of bargaining as follows:

> For the purposes of this section, to bargain collectively is the performance of the mutual obligation of the employer and the representative of the employees to meet at reasonable times and confer in good faith with respect to wages, hours, and other terms and conditions of employment, or the negotiation of an agreement or any question arising thereunder, and the execution of a written contract incorporating any agreement reached if requested by either party, but such obligation does not compel either party to agree to a proposal or require the making of a concession.[8]

As outlined above, bargaining issues are broken down into legal categories (*National Labor Relations Board v. Wooster Division of Borg-Warner Corp.*, 356 U.S. 342, 78 S. Ct. 718, 2 L. Ed. 2d 823 [1958]). It is via court rulings, State Labor Board rulings, and the NLRB rulings that the list will continue to add and subtract items from each of the three sections, and any list of items outlining the following three categories should never be considered final:

- Mandatory
- Permissive
- Prohibited

Mandatory bargaining issues deal with wage, hour, and other terms and conditions of employment. Many state collective bargaining laws require labor and management to bargain in good faith over mandatory issues. The law does not require an agreement; it requires only that the parties come to the bargaining table. It does not require a final contract.

Good-faith bargaining is a rather subjective concept. Generally, good-faith bargaining is when the two parties demonstrate a *sincere* and *honest* attempt to reach an agreement and present reasonable positions, tactics, and activities.[9] Per the Wagner Act (National Labor Relations Act [NLRA]), the employer cannot refuse to bargain with the labor union representing the employees; such a refusal would be considered an unfair labor practice.

In the spirit of the Wagner Act, many states have crafted laws that require the municipality to bargain with the labor unions, and if they refuse, it is considered an unfair labor practice. However, not all states have this level of protection. The Ninth Circuit Court, in *National Labor Relations Board v. Montgomery Ward & Co.*, stated that "the obligation of the parties to participate actively in the deliberations so as to indicate a present intention to find a basis for agreement, and a sincere effort must be made to reach a common ground."[10] Later, in *National Labor Relations Board v. Reed & Price Manufacturing Co.*, the courts indicated that bargaining in good faith implied a duty to bargain, or as noted in the case, "an open mind and a sincere desire to reach an agreement."[11]

Thus, by the 1940s, it appeared that the employer and the union must work to establish a final collective bargaining agreement. Since Congress felt that this section of the NLRA needed to be refined to clarify the spirit of the agreement, in 1947, the Taft-Hartley Act amended the NLRA to better define the obligation: specifically, to bargain in good faith "does not compel either party to agree to a proposal or require the making of a concession."[12]

NLRA § 8 defines employer unfair labor practices. Five types of conduct are illegal:

- Employer interferes with, restrains, or coerces employees in their pursuit of rights granted in section 7 (§ 8(a)(1))
- Employer dominates or interferes with the formation or administration of any labor organization or contribute financial or other support to it (§ 8(a)(2))

- Employer discriminates against employees who take part in union or collective activities (§ 8(a)(3))
- Employer discharges or otherwise discriminates against an employee because he/she has filed charges or cooperated with an NLRB action (§ 8(a)(4))
- Employer refuses to bargain in good faith with union representatives about wages, hours, and conditions of employment (§ 8(a)(5))

Student Activity

Research § 8 of the NLRA and identify the differences between § 8 and their individual state public sector labor laws pertaining to unfair labor practices.

1. Does your state have a section of law covering public sector unfair labor practices?
2. Compare and contrast your state unfair labor practices to the NLRA, § 8. How are they different? How are they the same?
3. What are the requirements for a union to file an unfair labor practice in your state?
4. What steps can labor and management take to prevent a charge of unfair labor practice?
5. Provide a detailed example of a public sector unfair labor practice for your state.
6. Research your state public sector labor agency/board and find a recent case that involves a charge of "unfair labor practice" and provide an overview of the case and what the two sides could have done to avoid a charge.

Students can review a copy of the NLRA at http://www.nlrb.gov/nlrb/shared_files/brochures/basicguide.pdf

If one party insists on including a mandatory item in the collective bargaining agreement, then the other party cannot refuse to discuss the item. As Holley, Jennings, and Wolters have pointed out, "management may not legally change current contract language governing an existing employment term involving a mandatory bargaining subject unless the current contract clearly gives management the right to do so; the union has clearly waived its right to bargain over the subject; or both parties have bargained in good faith and reached a voluntary agreement about the changes."[13]

When labor and management are bargaining over a mandatory item and they cannot reach an agreement, this is called a *impasse*. The impasse is when the two parties cannot come to a mutual agreement and are unwilling to mortify their positions and each feel that bargaining any further will not bring about any positive change or chance of establishing an agreement. In the event that a impasse is reached, the employer can implement its final offer. If a collective bargaining contract is currently in place, then management must wait until an impasse is reached before implementing the item, provided that the item has been bargained over with the union. In the event that a contract has expired, the employer must also meet the requirement of bargaining to impasse before implementing the item. Just because a contract has expired does not allow the two parties to change things covered in the contract without first bargaining over mandatory items. Because each individual state determines which items are considered mandatory, permissive, or prohibited, it is strongly recommended that you research your own state laws pertaining to collective bargaining for public employees—and specifically for firefighters.

Sample of Mandatory Subjects for Bargaining

- Wages
- Job classifications
- Evaluation procedures
- Grievance procedures (including binding arbitration of grievances)
- Health and safety matters
- Holidays
- Hours of hours
- Pension plan
- Insurance
- Funeral leave pay
- Vacation pay
- Overtime pay
- Seniority
- Shift differentials
- Staff reduction procedures
- Job promotion criteria and procedure
- Transfer procedures
- Clothing/tool allowance
- Call-back procedure/pay
- Management rights

In the event that one party wishes to bargain over a mandatory item and the other party refuses, this is referred to as an *unfair labor practice*. If two parties are unable to reach an agreement on mandatory items after meeting just once, it would not meet the standard for bargaining in good faith, so this would also be considered an unfair labor practice. Generally, meeting just once is not considered a good faith practice in the field of labor negations. Many state labor relation boards will rely on the *totality-of-conduct* concept, rather than a single action. Some of the following items that may be taken into consideration when the board examines a case of bargaining in good-faith: Did the two parties meet at reasonable times and places? Did the two parties exchange proposals and counterproposals? Did one of the parties refuse to bargain a mandatory item, did one side withhold information in an effort to prevent the other party from making a decision?[14]

The second category, *permissive* or nonmandatory items, comprises those issues that the parties may, but are not required to, bargain over. In this case, one party may bring forward an issue and it is up to the other party to decide if the items will be bargained. Each party must agree to bargain a permissible item. A permissive item cannot be bargained to a impasse. In the case of a permissive bargaining item, the party can unilaterally implement the change, and cannot be forced to bargain over the issue. Permissive bargaining items are the items that fall outside the realm of such mandatory items as wages, hours, and other terms and conditions of employment. Generally in the public and private sector, the union cannot take any type of economic action, such as strikes or picketing, to force bargaining over a permissible subject.[15]

Sample of Permissible Subjects for Bargaining

- Pension benefits for retired employees
- Determination and inclusion of supervisors in the contract
- Settlement of unfair labor charges
- Selection of bargaining team members
- Promotion of employees to supervisory positions
- Determination of amount of agency fee to be charged to nonunion members

The third category is *prohibited* bargaining items. The items that fall under this category are those that cannot be bargained over legally by either party, even if both sides agree. Each state will establish certain public sector laws that cannot be bargained because of state law. For example, the two sides cannot agree to hire only white male firefighters; this is protected under the Civil Rights Act. Another example is the two sides mutually agreeing that only current card-carrying union members will be hired by the department (a form of closed-shop clause). Many states have adopted the section under the 1947 Taft-Hartley provisions, preventing only current union members from being hired.[16]

Labor unions and employers have a legal duty to bargain in good faith. In some cases, the state government has required the two parties to come to the bargaining table but does not require the two parties to come to an agreement. Labor boards will often look at the totality of each party's conduct to determine if one of the parties has violated the duty to bargain in good faith. The board will often look at any overt behavior and any underlying motivation or intent by the two parties. This is not to say that a union or management could not take a firm stance on a bargaining issue. What is important is that the party must be willing to give and receive information about an offer at the table and be open to other offerings or discussions.

To come to the bargaining table with your first offer being a "take it for leave it" offer could cross the line into bad-faith bargaining. In the field of labor relations, this is referred to as Boulwarism, after General Electric Vice President Lemuel Boulware, who would come to the bargaining table with an offer considered to be "fair and firm" and backed up with research and documented facts. Management would argue that it did not entertain any other offerings because the union did not offer "new and convincing" information. As a result of the totality-of-conduct doctrine, it was determined that Boulwarism is bargaining in bad faith.[17]

Many states have adopted legislation that places a duty to supply information "relevant and necessary" in bargaining over wages, hours, and other conditions of employment. The municipality is often bound to supply, on request, information pertaining to the bargaining issues and to allow each party to bargain intelligently and effectively. For example, the State of Connecticut Municipal Employee Relations Act considers the duty to supply information as follows:

> Duty to Supply Information
>
> The duty to bargain in good faith extends to all labor management relations during the term of an agreement. An important aspect of this duty is the duty to furnish information. The obligation to bargain in good faith includes the obligation of both labor and management to provide relevant information that is necessary and relevant to the collective bargaining relationship. Wage or related economic information is presumptively relevant. In circumstances where a party claims information to be confidential, the Labor Board will apply a balancing approach which is adopted from the National Labor Relations Board case law.[18]

The wearing of union buttons and insignia can become controversial (fig. 4–2). In some states, this is a protected activity; however, as Carrel and Heavrin have argued, "this right is balanced against the employer's right to conduct business. If a button or insignia should, in particular circumstances, cause a disturbance, present a health hazard, distract workers, cause damage to product, or offend or distract customers, it may be prohibited."[19] The issue of placing union stickers on fire engines has raised a controversy, as some question whether it is appropriate to place union stickers on municipal vehicles. Further, some argue that wearing a metal union insignia on the uniform presents a safety issue, while others have said that the uniform has other items that are made of metal and, hence, that all metal should be removed from the uniform.

Fig. 4–2. The hot-button issue of union insignia. Depending on the state labor laws, the wearing of union buttons and insignia may be determined by management. Management may be able to establish an anti-adornment policy for nondepartment insignia.

Another issue over which labor and management disagree is the use of bulletin boards and meeting locations within the fire station. In general, the employees and labor union have no statutory right to use and/or display union information/material on the employer's bulletin board. In many cases, this has been addressed at the bargaining table and clearly defined in the collective bargaining agreement. For example, the Baltimore Fire Fighters, Local 734, Memorandum of Understanding states the following: "The Employer agrees to provide a bulletin board at least 36" × 48" labeled with the Union's name where notices of official Union matters may be posted by the Union. All Fire Department Bulletin Boards shall be used only for official notices of Union matters and Fire Department documents. Only the Chief of the Department has the right to order the removal of any literature not in compliance with the foregoing."[20]

The use of employer e-mail has also been called into question by labor unions. To what extent can the labor union use the employer's e-mail system for disseminating union material, notices, and so forth? For example, Clearwater, Florida, adopted a broad-brush policy limiting the union's use of the e-mail system. The labor board ruled, based on a series of past practices demonstrating that the union had been allowed to use the e-mail system, that the City of Clearwater must cease and desist applying an overly broad no-solicitation rule and allow employees to use the city's e-mail system to send union messages on nonwork time.[21]

Generally, a meeting with labor and management should be able to establish a middle ground to keep all parties happy when it comes to wearing union buttons on duty, use of bulletin boards in the station, and the use of the employer's e-mail system. In the event that an agreement cannot be reached, some states have adopted protected activities that would leave it up to the state labor board to rule on the disagreement. When these types of cases reach the labor board, they often have a long-term impact on the bargaining relationship.

From time to time, labor and management may have different opinions on how the collective bargaining agreement is written and the meaning behind the article. When labor and management cannot agree on the language in the contract and feel a violation has accrued, then a grievance is filed. Grievance administration is also a significant part of contract administration, as will be discussed in greater detail in chapter 6.

In the past 50 years many states have passed public sector labor laws allowing public sector employees the right to join unions and allowing unions to enter into collective bargaining. The NLRA excludes political subdivisions (municipalities, counties, fire districts, and the state) from the definition of "employer"; hence, each state is allowed to craft its own labor laws. Over the years, multiple attempts have been made in Congress to pass a national collective bargaining system for the public sector; as of today, no such law has been passed.[22]

Labor relations in the public sector has become a political process, and labor unions have invested a great deal of money behind the Democratic Party in an effort to win political support for organized labor. The level and significance of political power will differ from area to area and from union to union. When examining the labor relations process, one must take into account the political weight that is held by a given public sector union. During a local, state, or federal election, look around and see just how involved organized labor is in political campaigns. If they are not involved with political campaigns, then they are generally very active in lobbying activities. In any event, organized labor may ultimately need to call on the support of an elected official for passage of new labor laws, protection of existing labor laws, or anything else that may have an impact on their membership.[23] Politics can affect bargaining at the local level between labor and management.

No matter in how much detail states define the scope of bargaining in the public sector, questions will always be raised in the form of a legal challenge. In some cases, the legal challenge may be presented to the state labor board, asking for an opinion or the filing of an unfair labor practice, in which one of the parties is charging the other with refusing to bargain in good faith. In each of these cases, if one of the parties is still aggrieved by the decision offered by the labor board, then it can take the issue to court and ask for a opinion from a judge.[24] Through the rulings handed down by the courts and/or labor boards the scope of bargaining is refined. The responsibility falls to labor and management to stay current on their own state's scope of bargaining.

Bargaining power

Bargaining power can be defined as the amount of power each party has at the bargaining table. The Webbs explained that two forms are used to balance power: *method of collective bargaining* and *method of legal enactment*. The use of labor unions would provide additional power to the employees against the employer, providing more balance in power. In the early years, employees had two choices: to work or not to work; they did not have a say in the working conditions, and employers set the working conditions and wages and would often take advantage of employees. In the legal enactment form, the state and federal government could establish minimum standards for the employer-employee relationship. In this form, the widest protection could be provided to workers under the shield of labor standards.[25]

In the public sector, the employer agrees to deal exclusively with the labor union and to sit down and bargain over work-related issues, except in states that have outlawed collective bargaining in the public sector. This prevents the employer from dealing individually with each employee. This is a joint decision-making process. However, from time to time, the power balance may become uneven, tipping the scales in favor of one party. To begin with, once labor and management sign a contract, the contract will limit the public employer's discretion.[26] As noted earlier, the public sector unions show the greatest growth. As a result of this growing trend in the public sector, fire service administrators need to have a clear understanding and workings of labor relations. Many states have elected to create labor laws dealing directly with collective bargaining—and in some cases making it illegal. Some labor laws deal specifically with firefighters, law enforcement, or teachers.

Because states outlaw the use of strikes for firefighters, the law needed to also provide other means for labor and management to settle disagreements at the bargaining table. In the private sector, the union can take economic actions against the employer, such as going on strike, if both parties cannot come to terms during negotiations. In the case of binding collective negotiations, labor and management equally share the power, resulting in a legally binding contract.

In the event that labor and management cannot mutually agree on the terms of the contract, many public sector laws require *interest arbitration*, which is the final step when two sides have reached an impasse at the bargaining table. The question remains: is it best to designate a neutral third party to grant the demands, as opposed to having labor and management work out an agreement? There is always a certain degree of risk in subjecting to an arbitrator, as well as in not having any accountability for the award.

One option used to settle the disagreement is the use of *final-offer arbitration*, designed to increase the level of communication between labor and management at the bargaining table. As Bellante, Denholm, and Osorio have stated, "an arbitrator will never award a settlement that is less than management's final offer, so the union is guaranteed to obtain at least some of its demands, and will never come out worse than the status quo."[27] Lipsky and Katz have shown that more negotiated settlements are ratified using the final offer arbitration than traditional arbitration decisions.[28]

When one party believes that it has less bargaining power than the other, it tends to favor a interest arbitration that allows the arbitrator to decide the case, with greater scope of negotiation. When the two parties believe that the balance of power at the bargaining table is equal, they tend to favor regulations that support a greater opportunity to negotiate. By contrast, when the law only requires labor and management to meet and confer, the end result is that the two sides mutually agree on employment terms, and it is then documented in writing as a *memorandum of understanding*. However, in the event that the meet and confer does not result in a mutual agreement, there is no legal obligation to reach an agreement between the two sides.

As shown in the prior chapters, unions also hold a significant degree of political power at the local, state, and national levels. Public sector unions are some of the most powerful special-interest groups in the country. According to the Center for Response Politics, from 1989 to 2009, the American Federation of State, County and Municipal Employees donated nearly $40 million, and the National Education Association donated $28.4 million to political races; in each of these unions, 98% of the donations were given to the Democratic candidate.[29]

Public sector unions make strong lobbying efforts at all levels of government. As a result, public sector unions play a significant role in public policy in the United States. It is next to impossible to accurately measure the degree of power unions have in the public sector because it is not quantifiable.

At the local level, union endorsement and fiscal support are arguably the equivalent of a union hiring its own boss, who will be responsible for voting and approving future collective bargaining agreements. Elected public officials will often agree to support a collective bargaining agreement for a long-term contract, because it will keep the labor unions happy, and the fiscal reasonability will be held by someone else down the road. After all, elected officials need to gain public support to be reelected. In the private sector, a business must keep its stockholders happy and show a profit. The alternative for a private business that fails to make a profit is to go out of business. This is not the case with the public sector; it will always continue to collect taxes and, if needed, increase taxes. For example, if a municipality enters into a debate with a fire union or a police union, it is generally the only game in town, and the municipality does not want to endure the public's fury over a public sector strike—legal or illegal—that will disrupt service.

Now that the economy has taken a significant downturn, many elected officials and city/town administrators fear opening up the collective bargaining agreements, because of the political clout that labor unions wield at the local level. Unions will back elected officials and public administrators who support their goals: more public jobs, more organizing efforts for unions, and more taxes to pay for these programs. When municipal officials ask for contract concessions from fire and police unions—such as wage, health insurance, and other benefit reductions—unions often play the "fear factor" card—namely, that any reductions will result in a decrease in public safety.

Bargaining structure

Experts in the field of labor relations have said that there is a direct connection between the bargaining outcomes and the bargaining structure that is used in labor-management relations.[30,31] The bargaining structure involves a number of interconnected issues, including social, legal, administration, and economic factors.

Social factors can play a role in the bargaining outcome. Research by Van Kleef, De Dreu, and Manstead has indicated that negotiations taking place with a negative opponent often resulted in lower demands and with larger concessions compared to those with a nonemotional opponent.[32] A positive bargaining environment often will yield better results, and the two parties will feel that they have each made gains, which will subsequently play a role in the long-term labor relations between the two parties; this approach results in a win-win feeling for each party. By contrast, when negotiating with a negative opponent, the negotiator often turns to a more competitive approach, with very narrow concessions. Each side may request higher demands until the two sides are not even close to being in the same bargaining zone; this approach results in a win-lose feeling, with one party perceiving that it has lost and often results in an overall negative impact on the long-term labor relations process.

Legal limitations for collective bargaining can affect how labor and management approach the bargaining process. Keeping in mind that each state crafts its own public sector labor laws, the state laws may change from year to year, depending on who is in control of the government. In addition, some states require only that the two sides meet and confer; further, some states may not allow municipalities to enter into a collective bargaining agreement. In the latter situation, the union is somewhat at a disadvantage to demand certain items at the bargaining table. Labor and management may agree to a memorandum of understanding; nevertheless, the law limits the bargaining, and this shapes the bargaining structure. State labor board rulings, court rulings, state labor agencies, civil service rules, and legislative actions all have an impact on the nature and characteristics of the bargaining structure.

The administration of the municipality may change with the political tides, and this will have a direct impact on the bargaining structure. As discussed previously, many local unions will endorse a political candidate who represents the concerns of the union members. Once elected into office, it is expected that these elected officials will continue to support the union's demands at the bargaining table. In some cases, appointed officials (town managers, city managers, and town administrators), have short-term goals and are not interested in developing long-term labor relations. The appointed person may want to prove his/her worth to the appointing board by taking on the labor unions.

Some municipal leaders will want to stagger the bargaining cycle between the municipal unions, while other municipalities will want to bargain all of the contracts the same year to ensure that they can better plan the municipal budget. In some cases, there will be an overlap of the budget schedule and the final contract cost due to ongoing budgeting. In this case, the municipal budget may be approved with a projected expense for the pending collective bargaining agreement. Once the contract is agreed on, the municipal budget may need to be revised—and, in many cases, this is no easy task. Some municipalities will also require that any contract increases must be approved by the voters at an annual election or they may even call a special election.

Regardless of the type of government you work under, political forces do play a role in the bargaining process and the bargaining outcomes. The administration of the local union will also change from time to time. Some union presidents may be long term, while others are short term. Who does the bargaining for the union? While some unions may hire a labor attorney to serve as the chief negotiator, other unions elect a bargaining committee made up of the local union members, and still other unions rely on the national union to send a chief negotiator. Your bargaining structure should take into consideration the current administration of the municipality and the local union.

From an economic standpoint, with national unemployment at double digits, it is not in the current best interest of any union to demand a pay increase. Unions should be reviewing the tax revenue numbers of the community. Are the tax receipts above the projected expectations for the quarter or for the year? If the projected tax revenue receipts have been lower than expected, this means less money is coming into the municipality; less tax revenue in turn means less money to operate the municipalities, and that includes the cost of collective bargaining agreements.

Unlike the private sector, municipalities do not operate on a profit/loss balance sheet. However, most municipalities want a balanced budget and do not spend more than they are receiving via taxes, state aid, and other means. Some municipalities will strive for a budget that establishes a reserve surplus. The surplus is added to a special account to help the municipality in the event that the economy takes a sudden downward turn or in the event of unexpected expenses.

Labor unions should take an active part in the municipal budget process. Too few firefighters have any idea of the fiscal standing of their communities, but that's what determines the ability of the community to pay increases in the contract. In the public sector, budget administrators may hide needed funds within the budget, transfer funds from other areas, or even pad the line items to cover pending contract agreements costs. The process of contract negotiations has created a real challenge for municipal fiscal officers, who must prepare a budget 18–20 months in advance. Given that a majority of municipal budgets are allotted for wages, the fiscal accountability has a direct connection to the collective bargaining agreement, the negotiation process, and personnel management.

In some municipalities, the various local unions may establish a line of communication as to what has worked at the table and what has not worked. On the one hand, if several local unions band together, they can present a united front against the employer and demonstrate bargaining power. On the other hand, some unions will not want to share bargaining tactics with other unions within the community, as the municipal budget is only so big and each union is getting a piece of the pie; in other words, all the unions are in competition. For example, if the firefighters union were able to win a three-year contract with a 9% wage increase, why would they share the strategy that they used to win this wage increase, knowing that the police and teachers are also up for a contract? Although there tends to be some sharing of bargaining information and what articles have been agreed on, there remains information that is not shared between the unions.

Collectively, if all the unions in a municipality are looking for a significant change in health insurance, for example, this is a bargaining item that would benefit all employees within the municipality, so all the unions may band together and present a united front to bargain with the employer. This may be considered *coordinated bargaining* (also referred to as "coalition bargaining") when all or most of the unions are bargaining at the same table at the same time. Many municipal leaders are opposed to a coordinated bargaining session, believing that they will have less bargaining power. The goal is to pit unions against each other.

Coordinated bargaining is sometimes welcomed when the municipality has a master contract that covers multiple small bargaining units. This allows the main items to be placed in the master collective bargaining agreement, and then special items that pertain exclusively to a unit are outlined in an appendix. Items that may be unit specific include safety and health conditions, seniority requirements, work standards, shift bidding/changes/assignments, and determination of overtime distribution.[33] By streamlining the process, coordinated bargaining can be a time-saver for all the bargaining parties. No matter which process is utilized, it needs to work for both labor and management. It is up to the individual groups to determine which process works best for them in maintaining a positive bargaining process.

Another example of a bargaining process is called *pattern bargaining*. In the simplest terms, pattern bargaining is a method that is used when one collective bargaining is settled within a municipality and the municipal leaders and union leaders use the same agreement with all the remaining unions for a quick settlement and to minimize wage patterns across the municipality. The process of pattern bargaining takes one mutually agreed term and spreads it out across all the other labor unions in the municipality, which can be considered as an informal substitute for centralized bargaining. The goal is to minimize or avoid internal political problems between labor unions and management.[34] Marshall and Merlo research found that private sectors unions that used pattern bargaining resulted in higher payoff for union than other forms of bargaining, including sequential and simultaneous bargaining.[35]

Many unions and municipalities during the past couple of years have been faced with a poor economy marked by drastic budget reductions as have not been seen over the past 20 years. In turn, positions have been eliminated and workers have been laid off; some have even turned to mandatory furlough days. When the municipalities order a furlough day, all employees are given a day off from work without pay. This results in a saving in general wages for the municipality. However, the utilization of furlough days is controversial and, in some cases, is in violation of collective bargaining agreements. For example, the State of Oregon issued six furlough days during 2010 for most state agencies, in an effort to help offset the reduced revenues the state was receiving.

From 2009 to 2012, many municipalities and labor unions have entered into *concessionary bargaining*, to maintain the current workforce. Concessionary bargaining involves the municipality coming to the bargaining table and asking for *givebacks* from the labor union. For example, a fire department may be facing a reduction in line firefighter members based on the current collective bargaining agreement. In order for the employer to meet the terms and conditions as outlined in the collective bargaining agreement, it would have to reduce the number of firefighters to make up the projected budget shortfalls; the other option is to open up the current collective bargaining agreement and reduce some of the cost items contained within the agreement. Given that the majority of the fire department budget is made up of salaries, if the union gives back a pending pay raise as outlined in the contract, it could spare a number of projected layoffs. This spreads the loss across all members of the union, rather than those junior members who would be laid off.

Interestingly, some unions have taken a stand against concessionary bargaining, forcing the employer to make reductions in personnel in hopes that the public will not support the reductions. This move is placing a heavy emphasis on the political pressure of the public to force the municipal leaders to reverse the decision and reinstate the eliminated positions. Other labor unions have embraced wage and benefit givebacks as a means to save jobs.

On a positive note, with concessionary bargaining, unions could see a win for taking a concession on one item that the employer deeply needs (wages and benefits) in turn for receiving a larger role in management decision-making or reducing management rights. During these tight economic times, the traditional labor relations process may be altered by certain long-term consequences of short-term gains: For instance, if labor is willing to forfeit a wage increase today to help the current economic condition of the city, it may be rewarded down the road for cooperating with management; conversely, if labor unions take a strong stance and do not work with management, the labor-management relations may be damaged in the long term.[36] Nevertheless, elected officials may favor a short-term gain with labor unions, knowing that it may bolster the chances of being reelected, and the long-term consequences may fall on the next person to hold office.

Decentralized bargaining

Decentralized bargaining is common in public sector bargaining. In earlier years, the private sector favored a more centralized bargaining process that was used across a given industry or trade. Today, public sector bargaining is more decentralized, allowing each individual local to bargain specialized items and issues that pertain to its current employment. This approach places more control of the bargaining in the hands of the local union as compared with a national union bargaining for the majority of employees. For example, the National Association of Government Employee (NAGE) would bargain a master labor contract for all locals in a city/town/state and then allow each of the locals to add specialized items called sub-unit agreements as needed in a centralized bargaining process. The master agreement may cover items that will apply to all workers such as sick time, vacation time, dues check off, health and safety, representation recognition, grievance process, work schedule, etc. The master agreement saves the time and energy of each local union and the city/town/state from having to bargain over these issues repeatedly with each local union. The subunit agreements will all fall under the master agreement and may address specialized issues for each local union. In the decentralized process, the local union is allowed to address its own needs/wants in an individual collective bargaining agreement.

Decentralized bargaining will yield many different types of collective bargaining agreements across the land. Benefits and risks are associated with each bargaining process, and a primary concern with a decentralized process is that smaller, less powerful unions will not do as well at the bargaining table as they would with a more centralized process. Conversely, it can be argued that local unions will do better at the bargaining table under a decentralized process because they know the political culture and they personally know their bargaining partner.

The decentralized bargaining process takes into account the local market factors. In the decentralized process each party at the bargaining table will devise a structure that best works for it and maximizes returns at the bargaining table (Herman, 138). A municipality will be held accountable by the taxpayers who must shoulder the burden of any fiscal impact imposed by a contract change. For example, in a centralized process, the master contract could potentially break a local community that cannot absorb the fiscal burden placed on taxpayers by the master agreement. The taxpayers would show their displeasure at the polls by not reelecting these legislators.

All too often the bargaining process is viewed in terms of money—how much one side receives and how much the other gives up. The long-term focus in labor relations should be on building and maintaining a solid social relationship between the two parties. Many of the laws pertaining to labor relations are about what should not be done or what can be done. But there are no laws requiring that labor and management have a functioning relationship. The mixture of a dysfunctional bargaining relationship with most people's view of the bargaining process as a purely economic transaction leaves room for improvement. The viewpoint of the economic transaction can be traced back to early labor laws that required bargaining over wages.

Final Thoughts

This chapter has introduced bargaining terms for a collective bargaining agreement. Examples of the most common articles used in collective bargaining agreements have been provided. As you can see, collective bargaining agreements pertaining to firefighters have many similarities, even while they differ in certain terms and conditions. A collective bargaining agreement is often considered an individual agreement that pertains to the two parties and the terms and conditions that work best for each party. The collective bargaining agreement will affect how personnel management is performed.

Employees who are covered under a collective bargaining agreement have a right to belong to the labor union, as well as a right not to join the labor union. The collective bargaining agreement allows for an agency fee or user fee, allowing the union to assess members of the bargaining unit who do not join the union. The agency fee, established by the labor union for nonmembers, must be based on the cost of bargaining and administering the collective bargaining agreement. The agency fee cannot be used for political support or other nonbargaining items.

Many states have outlined items that are mandatory items for collective bargaining in the public sector. States may also outline items that are prohibited from collective bargaining. Each individual state will establish items that are mandatory and prohibited.

This chapter has explored the structure and theory of bargaining power as it pertains to public sector bargaining. A common theme so far in this book has been the role of politics in the collective bargaining process. This can be seen at the local level or at the national level when it comes to the political environment. Additionally, the structure that the bargaining process takes can also play a role in the overall process of collective bargaining. In many of the issues reviewed, it has become apparent that many factors come into play when examining the bargaining process in the public sector.

Key Terms

- Bargaining committee
- Bargaining structure
- Bargaining cycle
- Boulwarism
- Chief negotiator
- Coalition bargaining
- Collective bargaining agreement (CBA)
- Concessionary bargaining
- Confidential employee
- Coordinated bargaining
- Decentralized bargaining
- Dues checkoff
- Duty to supply information
- Exclusive bargaining representative
- Final-offer arbitration
- Givebacks
- Good-faith bargaining
- Impasse
- Interest arbitration
- Management rights
- Mandatory bargaining
- Memorandum of understanding (MOU)
- Method of collective bargaining
- Method of legal enactment
- Preamble
- Recognition clause
- Pattern bargaining
- Permissive bargaining items
- Prohibited bargaining items
- Supervisory employee
- Taft-Hartley Act
- Terms of employment
- Totality-of-conduct
- Unfair labor practice
- Union security
- Wagner Act

Exploring Issues on the Web

City of Austin (Tex.) and Local 975 Collective Bargaining Agreement: http://www.ci.austin.tx.us/fire/downloads/finalcbaw_appendices.pdf

City of Bloomington (Ind.) and Local 3715 Collective Bargaining Agreement: http://bloomington.in.gov/code/level2/T2_C2.34.html

City of Seattle (Wash.) and Local 27 Collective Bargaining Agreement: http://www.cityofseattle.net/personnel/resources/pubs/Local_27_2009-11_Final.pdf

Pennsylvania Police and Fire Collective Bargaining Law Act 111: http://www.portal.state.pa.us/portal/server.pt?open=514&objID=552988&mode=2

References

1. Carrell, M., and C. Heavrin. 2010. *Labor Relations and Collective Bargaining*. 9th ed. Upper Saddle River, N.J.: Pearson Prentice Hall. P. 107.
2. Montana State Law, Title 39, Chapter 31-103(11)(a). http://law.justia.com/montana/codes/39/39-31-103.html (accessed September 1, 2010).
3. Holley, W., and K. Jennings, and R. Wolters. 2009. *The Labor Relations Process*. 9th ed. Mason, Ohio: South-Western. P. 162.
4. Sloane, A., and F. Witney. 2010. *Labor Relations*. 13th ed. Upper Saddle River, N.J.: Prentice Hall. P. 340.
5. Pynes, J. 2007. What public employee relation boards and the courts are deciding. In *Public Personnel Administration and Labor Relations*, ed. N. Riccucci. Armonk, N.Y.: M.E. Sharpe. P. 432
6. Carrell and Heavrin 2010, p. 113.
7. Aitchison, W. 2010. *The Rights of Firefighters*. 4th ed. Portland, Ore.: LRIS Publications. P. 14.
8. Taft-Hartley Act, § 8(d).
9. Holley et al. 2009, p. 268.
10. *National Labor Relations Board v. Montgomery Ward & Co*. 1943.
11. *National Labor Relations Board v. Reed & Price Manufacturing Co*., 118F.2d at 875.
12. Taft-Hartley Act, 29 U.S.C.A. § 141 et seq.
13. Holley et al. 2009, p. 268.
14. Herman, E. E. 1998. *Collective Bargaining and Labor Relations*. 4th ed. Upper Saddle River, N.J.: Prentice Hall. P. 175.
15. Katz, H., and T. Kochan. 2004. *An Introduction to Collective Bargaining and Industrial Relations*. 3rd ed. New York: McGraw-Hill. P. 55.
16. Fossum, J. 2009. *Labor Relations*. 3rd ed. New York: McGraw-Hill Irwin. P. 217.
17. Holley et al. 2009, p. 270.
18. Connecticut Department of Labor, Project Management Office. 2010. Municipal Employee Relations Act. Conn. Gen. Stat. § 7-467, et seq. http://www.ctdol.state.ct.us/csblr/mera7-467.htm (accessed August 1, 2010).
19. Carrell and Heavrin 2010, p. 163.
20. Baltimore Fire Fighters, Local 734. 2003. Baltimore Fire Fighters Memorandum of understanding. Article 10. http://www.dol.gov/olms/regs/compliance/cba/pdf/cbrp0249.pdf (accessed August 1, 2010).
21. State of Florida Public Employee Relation Commission. 2006. *Local 1158, Clearwater Fire Fighters v. City of Clearwater*. Case No. CA-2005-086. http://www.iaffdistrict12.org/ULP_3.pdf (accessed August 1, 2010).
22. Aitchison 2010, p. 6.
23. Brock, J., and D. Lipsky. 2002. *Going Public: The Role of Labor-Management Relations in Delivering Quality Government Services*. Champaign, Ill.: Industrial Relations Research Association.

24. Gershenfeld, W., and G. Gershenfeld. 1994. The scope of collective bargaining. In *Handbook on Public Personnel Administration and Labor Relations*, ed. J. Rabin, T. Vocino, B. Hildreth, and G. Miller. New York: Marcel Dekker. P. 56.

25. Webb, S., and B. Webb. 1897. *Industrial Democracry*. London: Longman, Green.

26. Nigro, F. 2007. Collective negotiations in the public sector. In *Public Personnel Administration and Labor Relations*, ed. N. Riccucci. Armonk, N.Y.: M.E. Sharpe. P. 265.

27. Bellante, D., D. Denholm, and I. Osorio. 2009. Why public sector unionism is a bad deal for taxpayers and representative government. *Policy Analysis*. September 28 (645). http://www.cato.org/pubs/pas/pa645.pdf (accessed July 20, 2010). P. 16.

28. Lipsky, D., and H. Katz. 2006. Alternative approaches to interest arbitration: Lessons from New York City. *Public Personnel Management*. 35 (4): 265–281.

29. Bellante et al. 2009, p. 10.

30. Hendricks, W., and L. Kahs. 1982. The determinants of bargaining structure in U.S. manufacturing industries. *Industrial and Labor Relations Review*. 35 (2): 181–95.

31. Seo, Y. 2008. The influence of bargaining structure and positive affect on decision making in negotiation. Paper presented at the IACM 21st Annual Conference, Chicago, Il, July 3–6, 2008. http://ssrn.com/abstract=1298571 (accessed Aug 1, 2010).

32. Van Kleef, G. A., C. K. W. De Dreu, and A. S. R. Manstead. 2004. The interpersonal effects of anger and happiness in negotiations. *Journal of personality and social psychology*. 86 (1): 57–76.

33. Katz and Kochan 2004, p. 174.

34. Fossum 2009, p. 242.

35. Marshall, R., and A. Merlo. 1996. Pattern bargaining. Federal Reserve Bank of Minneapolis, Research Department staff report 220. http://www.minneapolisfed.org/research/sr/sr220.pdf (accessed August 1, 2010).

36. Becker, B. 1988. Concession bargaining: The meaning of union gains. *Academy of Management Journal*. 31 (2): 377–87.

NEGOTIATION PROCESS AND IMPASSE

5

Learning Objectives

After reading this chapter, you should be able to:

- List negotiating skills.
- Recognize the importance of establishing a bargaining team.
- Explain the steps in establishing a bargaining process.
- Recognize the importance of costing a collective bargaining agreement.
- List the steps labor and management can take in the event of a bargaining impasse.

Negotiation Process and Impasses

This chapter examines the process that labor and management often take to negotiate a collective bargaining agreement and the steps that can be taken to prevent a negotiation impasse during the process.

This chapter will briefly review the negotiation process. A majority of people reading this text will not be involved in the actual bargaining process; nevertheless, it is important to understand what takes place during the bargaining process. The study of negotiation skills could comprise a book in itself. The intent of the section is to briefly review the bargaining process to expose the reader to some of the steps labor and management will often take in the event of a bargaining impasse.

The first order of business is to explore how bargaining teams are established, along with the roles members play within the team. This chapter will touch on the skills needed by a labor relations negotiator. This chapter will explore the development of a bargaining proposal and what is often considered in preparing a proposal and in establishing the bargaining-session ground rules. Consideration will be given to contract costing and the importance of accurately costing a labor contract. Miscalculations for a collective bargaining agreement can be very expensive and long lasting. When drafting collective bargaining agreement language, each party must consider the spirit and the intent of the article to ensure that it is articulated in the contract verbiage. Often, ambiguity in contract language is used as part of the process or inadvertently. Labor and management should determine the risk of using ambiguous language in the contract. As you will see, ambiguous language may be needed in order to move the bargaining process along and to ratify a collective bargaining agreement.

Bargaining entails each party giving and taking to reach a mutually acceptable agreement. However, in some cases, no matter how much giving and taking is offered, agreement cannot be reached. When labor and management cannot come to terms for a contract, this is called a bargaining impasse. As long as there are negotiations, there is always the risk of an impasse. Each side should identify in advance, at the pre-negotiating stage, steps that could be taken in the event that negotiators cannot resolve their differences.

Bargaining Teams

The process of labor negotiations involves people; consequently, emotions and egos can delay the signing of contract or lead to a negotiation impasse. People generally have some level of experience at negotiations, be it good or bad: A person may have negotiated a deal to buy a new car with the car dealer; a person may have negotiated a contract to buy a house; a parent may have negotiated a time for a child to be home. Depending on the negotiations, it may be a one-on-one process or it may involve several people on each side. Labor negotiations often involve multiple people sitting at the bargaining table. Labor and management each have a goal going into the bargaining process, to bring a tentative agreement back to the stakeholders (the union must take the tentative contract back to the membership for ratification; the management side of the table will often have to have the final contracted reviewed by the legal department and approved by the municipal officials). In the public sector, management will need to determine whether it will appoint a team of managers from within the municipality or hire an outside negotiator (a labor attorney or a professional negotiator). The firefighters' union will often have to make the same decision regarding whether it will serve as the negotiating team, hire a labor attorney, or ask for assistance from the national union representing the local.

Labor and management must each decide how much authority to delegate to the negotiators. In most cases, the two sides will be required to bring a tentative labor agreement back to the larger management team and the union membership for a vote to ratify the contract. Therefore, it is critical that each side select a spokesperson who will represent the team at the bargaining table and who will be responsible for coordinating and managing the bargaining team during the process.

The spokesperson will be charged with delivering concession items, identifying and communicating team priorities to the other party, and delivering threats and follow-up comments.[1]

The makeup of the bargaining team will often depend on the current labor relationship and the size of the municipality and labor union. In smaller unions, it is common for three to four people to represent the union at the bargaining table. If needed, the local union may call on the international union for assistance in a particular area, such as negotiations, economics, benefits, and specialized items being bargained over. The local union may allow the negotiator from the international union to conduct all the bargaining and present the local with a tentative collective bargaining agreement.

Negotiating Skills

- Must communicate well orally and in writing
- Should be an active listener and not dominate the conversation
- Must be able to understand their position and, equally important, the opposing side's position
- Must be trustworthy
- Does not allow personalities to interfere with the process
- Must be well organized and detailed

What makes a difference is how the bargaining session progresses and the end results. As discussed in chapter 4, labor relations is about establishing a long-term bargaining relationship. By contrast, buying a car or a house, instead of being about establishing a bargaining relationship, is about walking away with the best possible deal; in some cases, it may end with a "drop-dead offer," carrying no chance of further discussion or offers. Knowing when to walk away is important before entering the bargaining process as a consumer, since it is not about how the other side looks at you when the deal is done; the bargaining process is about a business move that allows you to retain the best possible value in the process. By contrast, walking away from an offer does not yield the best results in labor negotiations. In labor negotiations, the idea is to build trust and respect and to create value so that a long-term bargaining relationship can be maintained.

One common mistake people make is that they may think they are a great negotiator because of the deal they have secured when buying a car or a house. There are many approaches to negotiating, and labor negotiations require a unique set of skills and techniques as compared to bargaining in a short-term relationship or negotiating a deal over material goods. At the bargaining table, when it comes to securing a contract, the negotiator's actions often speak louder than the words spoken; hence, rather than just talking about building trust and respect, let your actions show that you want to build the mutual trust and respect.

There are many seminars and classes on effective negotiations, and enhancing one's negotiation skills through such classes is recommended before sitting at the bargaining table to negotiate a labor contract. A lot is at risk in negotiations on behalf of the entire union body, as compared to negotiating for yourself. Negotiators who represent management have been terminated or demoted because they made a costly mistake during the bargaining process or if it appeared that they were giving the union the keys to the castle.

As a chief negotiator for labor or management, you will be held accountable for your results. Mistakes could be very costly immediately or down the road in the collectible bargaining agreement. A common rule of thumb is the *80/20 rule*, which specifies that you should spend 80% of your time and energy in preparing to negotiate and about 20% of your time and energy in actually performing the negotiations (fig. 5–1). When each side or just one side enters the bargaining process with a lack of data/information, it tends to prolong the bargaining process and, because of the lack of information and preparation, results in the inability to articulate a proper response.

Fig. 5–1. The 80/20 rule. Each bargaining team must spend time preparing for the bargaining session. For examples, research current trends, comparable wages and benefits, pending legislation, and past grievances/arbitration awards.

From management's perspective, the question is often asked, should the fire chief be the chief negotiator at the bargaining table? Some have argued that playing such an integral part in the bargaining will have a direct impact on the fire chief's ability to lead the department. In a majority of cases, the fire chief will sit in on the bargaining session to act as a resource for the chief negotiator, along with the director of human resources and the finance director. In the case of fire and law enforcement agencies, one tends to see the chief of department serving as a resource because of his/her specialized knowledge in the operation of the department and the proposed contract changes.[2] In other cases, the fire chief is an active member of the bargaining committee with positive results. Each individual community will need to determine if it is best for the fire chief to be an active member of the bargaining committee.

The makeup of the management team will depend on the municipality. However, when the fire chief is part of the bargaining team and things are not going well, the rank and file may feel that the chief is to blame. In the event that the fire chief is not at the bargaining table, the members may feel that the chief is not concerned or lacks the administration to bargain. No matter who makes up the bargaining team, one can quickly list the pros and cons of any selection.

A majority of municipalities will select the city manager or chief administrator to serve as the chief negotiator. If this is an appointed or elected position, then the negotiation sessions may be affected by local politics, especially common when the union endorses a certain candidate for the position of mayor. Once the new mayor has been voted into office, it is expected that the mayor will return the favor for the union by lightly bargaining. If the mayor is a one-term elected mayor, the person may be inclined to give the union what it wants to keep members happy and let the next mayor worry about the issues approved that day. To take this one step further, the newly elected mayor, whom the union supported, may appoint a new fire chief. Now sitting at the bargaining table will be two more people who may have some allegiance to the labor union. The new mayor and the new fire chief may know that their time in office will be short lived, and to make things run smoothly, they may be inclined to go easy on bargaining.

Some municipalities will use an outside legal counsel as the chief negotiator. In the event that talks turn negative, the pointing of fingers can be directed at the outside person, relieving some of the political pressure. However, there are also risks when a municipality uses outside negotiators at the bargaining table. For instance, the outside negotiators may be less concerned about maintaining the bargaining relationship than about how hard they bargained and the end results they achieved even at the risk of damaging the bargaining relationship.

Negotiations experts report that bargaining teams that have experienced an impasse during bargaining run a greater risk of experiencing another impasse at the next bargaining session or of the session's ending in lower gains (lose-lose). When a bargaining team has a history of impasses during negotiations, this tends to follow them in further negotiations.[3] This further drives home the issue that labor relations is about long-term relationships. What is said and done at the bargaining table will have a long-lasting impact on the relationship.

Bargaining Proposal

Before entering the formal bargaining session, labor and management should list the items that they deem important enough to be the focus of the bargaining process. The first line of review often rests in the evaluation of the current collective bargaining agreement. When labor and management attempt to alter the contract and how it is administered, this is referred to as *fractional bargaining*.

Labor and management will evaluate the current contract to determine the perceived weaknesses and strengths from each point of view. For example, has one article been the subject of a majority of grievances over the life of the current contract? Areas of the contract representing a high number of grievances should be considered topics for discussion at the bargaining table, with the goal of determining the root cause of the grievances. (Are the grievances perhaps the result of poorly written contract articles?) In addition, each side should closely examine current articles that have been the subject of complaints or misunderstandings, to determine the appropriate action that can be taken to minimize the complaints or misunderstanding in the proposed contract. These areas of complaint and misunderstanding may result in the filing of grievances down the road, so being proactive in addressing pending issues in the current contract may be of mutual benefit to labor and management.

One inherit limitation in the evaluation of grievances is that labor unions will sometimes solicit grievances to file against management. A grievance can be filed under the idea that the union believes that a violation of the contract has occurred; thus, in preparing the grievance, the union does not have to show that an actual violation has occurred, only that it believes a violation has occurred. It has been my experience that unions will use this tactic to demonstrate to management that they are unhappy, since management has a responsibility to respond to the grievances. The union may also file a high number of grievances on a specific article to advance it to the bargaining table later on, regardless of whether the grievances have merit. So in this instance, determining the area of high grievance filings may not be an accurate depiction of the contract language.

Labor unions may conduct a survey of its members to determine concerns that the general members have and issues that they feel should take priority at the bargaining table. This allows the individual members of the union a voice in the bargaining direction. After all, one of the reasons people join unions is because otherwise they would not have a voice in the job.

Another area that the union and management may examine together is industrywide issues. What are some of the major issues currently being discussed at the bargaining table pertaining to firefighters? Collectively, there is an attack on health benefits: management wants to pay less in the increase of health care, while labor wants to pay less for health care. Because this issue affects more than just one local union, it may be best to coordinate resources, to arrive at a successful agreement. Resources may be pooled at the state level or the international level to aid local unions with negotiating industrywide issues.

Generally, the labor union will present a list of items it would like to bargain over with management. Management will compare the union's list to its own; hence, it is equally important for management to properly research the items that it wishes to bargain over (fig. 5–2). Several meetings may be necessary to produce a final, mutually agreed on bargaining list.

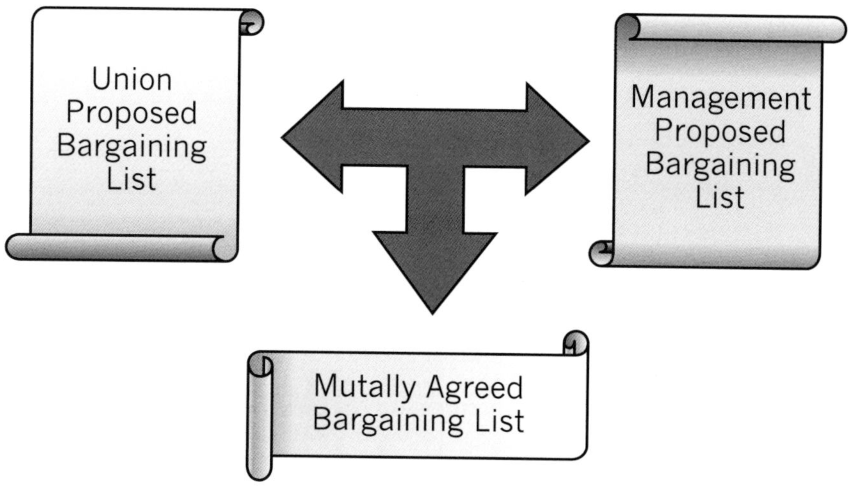

Fig. 5–2. Mutual bargaining list. Labor and management will establish a list of items each side would like to be discussed during the bargaining session. Labor and management will need to narrow down the items to be discussed and agree on the items to be discussed.

Management will often utilize senior fire officers on issues that management should be bargaining over (providing that the senior fire officers are in a different union or are not covered by a union). Management will involve the municipal human resource office in conducting research and presenting data pertaining to wages, benefits, and so forth. Management may also reach out to the legal department to see if any recent rulings will impact the contract or areas with which management should be concerned. Elective municipal officers may also play a role in the preparing items to be negotiated.

Management will often have various groups with conflicting interests, so the bargaining team must balance these interests against the pressure being placed on the team. For example, during my tenure as fire chief, the board of selectmen provided me with a direction it wanted to see the fire department move. Hence, the selectmen's consideration was given a great deal of weight. One area was the issue of overtime and sick-time abuse. After extensive consultation with the union in a partnership that each side could live with, a plan of action was implemented. At the time, the elected officials were pleased with the decrease in spending and the fiscal control implemented. I had to balance my professional ideas/direction with those I am accountable to. Considering that the board of selectmen is an elected board, the interests and personalities differed significantly from one election to the next. As the board changed, attempts to reconcile any differences seemed to fail, and the relationship became counterproductive. On a political front, the dynamic nature of politics can play into bargaining and the operation of the fire department over a short period of time.

Proposal assessment entails the following steps:

- Assessment of the other party
- Recognition of current trends in bargaining in the area fire departments
- Knowledge of recent settlement agreements
- Incorporation of previously negotiated items
- Assessment of prior grievances (any trends)
- Listening to stakeholders' current concerns/feedback
- Assessment of risk
- Assessment of the situation
- Identification of alternatives for bargaining items
- Assessment the legality of items

Information/data collection is critical to preparing any proposal for bargaining. Management and unions will need to secure information to prepare for bargaining. Data must be collected from more than one source if possible. With any type of information collection, availability and reliability is a concern. In the public sector, management and labor will often have public fiscal reports available to them. In the private sector, labor unions will often face more of a challenge in securing fiscal data, industry statistics, profits, sales, and industry trends from the company.

Before the bargaining session opens, each team will want to discuss the strategies it plans to use at the bargaining table. Bargaining strategies tend to go hand-in-hand with bargaining power. The old-fashioned approach, to go in strong and take as much as you can, no longer works because bargaining is not merely about who ends up with the larger piece of the fiscal pie; nevertheless, some people still use this outdated strategy. Today, one strategy used to negotiate is reflective of the needs for each party at the bargaining table; rather than one best strategy, the bargaining team may use several strategies throughout the entire process. Different items being discussed, on different days, may all require different strategies.

One final strategy should be discussed as a team: In the event of an impasse, how will we proceed and what steps can be taken to prevent harm to the bargaining relationship? At some point, the two chief negotiators may consult outside the bargaining room. Many contract disputes have been resolved by two people taking a short walk and talking it out.

Negotiation session

The first order of business at the opening negotiation session is to establish the ground rules. The two parties will determine where, when, and how long they will meet. Sometimes the two sides will also place a self-imposed deadline for agreeing on a contract. If the labor union is newly formed, the two parties may discuss the size and makeup of the bargaining team. In this case, each side will mutually agree to who will be its representatives and will ensure that only these members will be involved in the bargaining process; changing bargaining team members during the process would leave a team at a disadvantage. It is at this point that each party will identify a *chief negotiator*. Given a well-established bargaining relationship, establishment of the ground rules and identification of the chief negotiator may not take much time.

Generally, at the first meeting, the two parties may exchange proposals for items to be discussed at the bargaining table. In some cases, the union will present its proposal first to management, and at the second meeting, management will offer a counterproposal. Again, there is no set right or wrong way to proceed; rather, it is about what works best for the two parties.

As Malhotra and Bazerman have observed, "negotiators will often aim to build strong coalitions to weaken the other party's alternatives."[4] The history between the two parties will often set the stage for the bargaining process. The first couple of bargaining sessions may start off with light-hearted conversation, hopefully setting a positive tone for the following sessions. However, this is not always the case, and labor and management may enter the bargaining stage with an "us against them" approach that does not serve the bargaining process well.

Once the two parties have received and reviewed the proposals and counterproposals, the two sides will attempt to break down the items into economic and noneconomic items. The housekeeping items can be quickly agreed on and removed from the table. Another approach is to identify the noneconomic and less important issues and attempt to settle these items first, leaving the economic items on the table. There is a debate as to whether total removal of items from the bargaining table at this early stage is best. In some cases, the two parties will mutually agree to remove the items, with the understanding that they can be brought back to the table for further discussion and negotiating. Generally, if the two sides agree to total removal of an item from the bargaining table, then they are bound to the agreement to remove the item. How items will be removed from the bargaining table should be discussed and agreed on at the opening session to prevent misunderstandings.

One of the biggest challenges each party will face before negotiations truly begin is the bargaining priorities, strategies, tactics, and how to accept a tentative agreement. The next challenge facing each side is the release of information during the bargaining process. Some states have labor laws that allow the two parties to close the bargaining session to the public; other states have laws that exclude bargaining sessions from open-meeting laws; and still other state laws are silent on the issue, so the bargaining session could be open to the public. Today, more than ever, a push for transparency within government is having an impact on bargaining sessions. The public is now demanding to know what is being bargained at the table. Taxpayers, who foot the bill for the decisions being made behind closed doors, are waging considerable pressure on governmental officials to make the bargaining process more transparent.

Division remains as to whether the two sides should release any information during the process. One option is to offer a joint press release to the media during prolonged negotiation sessions. By providing status updates, rumors and speculations about how things are progressing can be minimized. Conversely, each side may agree that no information is to be released to anyone outside the bargaining room. In this case, union members may feel isolated from the bargaining process and that they have little say in how the team should be proceeding; each party must place a great deal of trust in its negotiation team. Only when the two parties have reached a tentative agreement will the team provide an explanation of the changes in the collective bargaining agreement. It is at this point that the two teams must sell the tentative contract to the stakeholders; hence, the bargaining team's job is not complete just because there is a tentative agreement. If the stakeholders vote down the agreement, the bargaining team will need to go back to the bargaining table to renegotiate.

The following states currently have laws allowing the government to close bargaining sessions to the public:

- Hawaii: Hawaii Rev. Stat. § 92-5(a)(3)
- Illinois: 5 Ill. Comp. Stat. 120/2(c)(2)
- Louisiana: La. Rev. Stat. Ann. § 42:6.1(A)(2)
- Maryland: Md. Code Ann., State Gov't § 10-508(a)(9)
- Massachusetts: Mass. Gen. Laws Ann. Ch. 30A, § 11A 1/2(3)
- New Jersey: N.J. Stat. Ann. § 10:4-12(b)(4)
- New York: N.Y. Public Office Law § 105(1)(e)
- Ohio: Ohio Rev. Code Ann. § 121.22(G)(4)
- Pennsylvania: 65 Pa. Cons. Stat. Ann. § 708(a)(2)
- Rhode Island: R.I. Gen. Laws § 42-46-5(a)(2)
- South Dakota: S.D. Codified Laws § 1-25-2(4)
- Vermont: Vt. Stat. Ann. Tit. 1, § 313(a)(1)

The following states have either exempted themselves from the open-meetings laws or defined their laws in such a way that they do not apply to public sector bargaining:[5]

- Connecticut: Conn. Gen. Stat. § 1-200(2)
- Indiana: Ind. Code § 5-14-1.5-2(b)(3)
- Iowa: Iowa Code Ann. § 20.17(3)
- Kentucky: Ky. Rev. Stat. Ann. § 61.810(1)(e)
- Nevada: Nev. Rev. Stat. 288.220
- New Hampshire: N.H. Rev. Stat. Ann. § 91-A:2(I)(b)
- New Mexico: N.M. Stat. Ann. § 10-15-1(H)(5)

- Washington: Wash. Rev. Code § 42.30.140(4)
- Wisconsin: Wis. Stat. § 19.82(1)

As a general rule, it is best for the union to offer the first press release to the public. The members of the union will want to see the bargaining team announce that it has reached a tentative agreement and has a good contract. If management announces first that it has secured a good contract, it leaves the impression that management is the winner in the bargaining process. Even though this may seem like a minor issue, that impression will play a significant role in how the union's rank and file will vote. If union members have the impression that management is taking advantage of them, they will vote the tentative contract down.

Contract costing

One issue critical to collective bargaining is the fiscal cost for the contract agreements. Each article in the collective bargaining agreement, be it from an economic or non-economic standpoint, will have some cost impact. At the bargaining table it is common to hear that a given article is a *no-cost* item. At some level, a cost is always attached to a given contract article, so the true economic cost must be determined. Granted, it may seem like a no-cost item this year, but what will it cost the municipality down the road? When reviewing bargaining articles, look not only at the current contract economic impact but also at long-term costing. For each cost change made, negative or positive, each side will need to constantly reevaluate the cost of the contract. One minor miscalculation at the bargaining table may result in a significant fiscal impact for years to come. Mistakes made at the bargaining table are generally very difficult to fix once they have been ratified in a contract. For example, if you sign a three-year contract, it may be another three years before this mistake could be addressed—and at what cost, since the other party will demand something in turn for correcting the mistake. A running tally should be monitored from each side of the table. The union will want to monitor the costing of the contract to determine the value of management's proposal. Management will want to monitor the contract costing to ensure that they are not overspending.[6]

Fire department personnel costs make up the largest part of the fire department budget. As a result, labor cost in the contract must be carefully calculated. A small change in labor costs could have a significant impact on the personnel budget for years to come. In order for management to control cost, it must have a firm understanding of the contract, and how those expenditures can impact the bottom line of the budget. Management will often accept or reject a proposal on the basis of the total cost and impact to the collective bargaining agreement. At the bargaining table, employee benefits are generally divided into two main sections:

- *Time not worked*, encompassing vacation, holidays, sick, funeral leave, union business
- *Security/health/life*, encompassing health insurance, life insurance, pension obligations, social security

The process of costing fringe benefits (compensation in addition to direct hourly wages) can be challenging, as this may be the most detailed area of most labor contracts. The third area of concern (after time not worked and security/health/life) comprises items that do not directly affect compensation yet play a role in the fiscal impact for the life of the contract, such as training programs, work assignments, additional specialty certifications, education, and seniority rules.[7] Because these are not so easily quantified, employees often forget about the hidden cost that the employer must pay—for example, the matching portion of a pension plan, the health/dental plan, and benefits for time not worked. To remind employees of the cost of such benefits, many communities have printed on the paycheck stub the true cost of the benefits.

Many human resource departments have prepared information sheets breaking down the true cost of benefits the municipality offers to the employees, including what the employee pays and what the employer pays. Figure 5–3 shows a detailed list of the cost of insurance prepared by the Town of Lexington, Massachusetts. Note that the employees have a menu of health/dental plans to choose from, and each plan is broken down to show actual costs. The top of figure 5–3 shows the town contribution, the employee contribution, and the total cost.

Personnel cost entails not only wages but also the overall compensation package, including the component parts. It is important to keep in mind that any increases in the area of time not worked often have hidden costs. For example, the union may argue that the employee is being paid a salary, and allowing him/her to attend outside training as part of the work schedule does not affect the department. Yet, the department may be faced with back-filling this position with an overtime employee, based on minimum shift requirements in the contract.

Generally, contract costs are considered fixed costs and adjustable costs. In the case of a fixed cost, the firefighter may be given a Haz-Mat stipend each July 1 in the amount of $500.00. This is a fixed cost occurring each year at the same rate and based on the number of employees it impacts. Assuming a Haz-Mat stipend at 2% of the firefighters base pay, the adjustable cost of the stipend will change depending on the base pay as specified in another section of the contract. Management will need to determine the cost of the increase for the base pay and how this additional increase in base pay will affect the percentage of the Haz-Mat stipend. Unions generally prefer a percentage indicator in the contract so that pay will increase based on other factors, while management prefers fixed costs. For example, if a contract has a number of items that are adjustable with the base pay, a 2% increase in base pay might amount to a total increase of 3.5% as a result of the other items that will be affected by the increase. On the one hand, management would be mistaken if it took the total payroll and multiplied it by the 2% in this example, because the actual cost of the increase is 3.5%; on the other hand, employees often focus on the base pay increase (2%) and do not recognize the entire value of the increase (3.5%).

Fiscal Year 2010 Town of Lexington (MA) Health Rates -- Effective 7/1/2009 to 6/30/2010

	Town Annual Cost	Employee Annual Cost	Total Annual Cost
Family Plan: **$25,415.16**	$ 20,332.13	$ 5,083.03	$ 25,415.16
Family HMO Plan: **$18,934.44**	$ 16,094.27	$ 2,840.17	$ 18,934.44
Individual Plan: **$9,731.40**	$ 7,785.12	$ 1,946.28	$ 9,731.40
Individual HMO Plan: **$7,276.44**	$ 6,184.97	$ 1,091.47	$ 7,276.44
Retiree/Senior Plans Family: **$26,663.64**	$ 21,330.91	$ 5,332.73	$ 26,663.64
Retiree/Senior Plans Individual: **$11,230.56**	$ 8,984.45	$ 2,246.11	$ 11,230.56

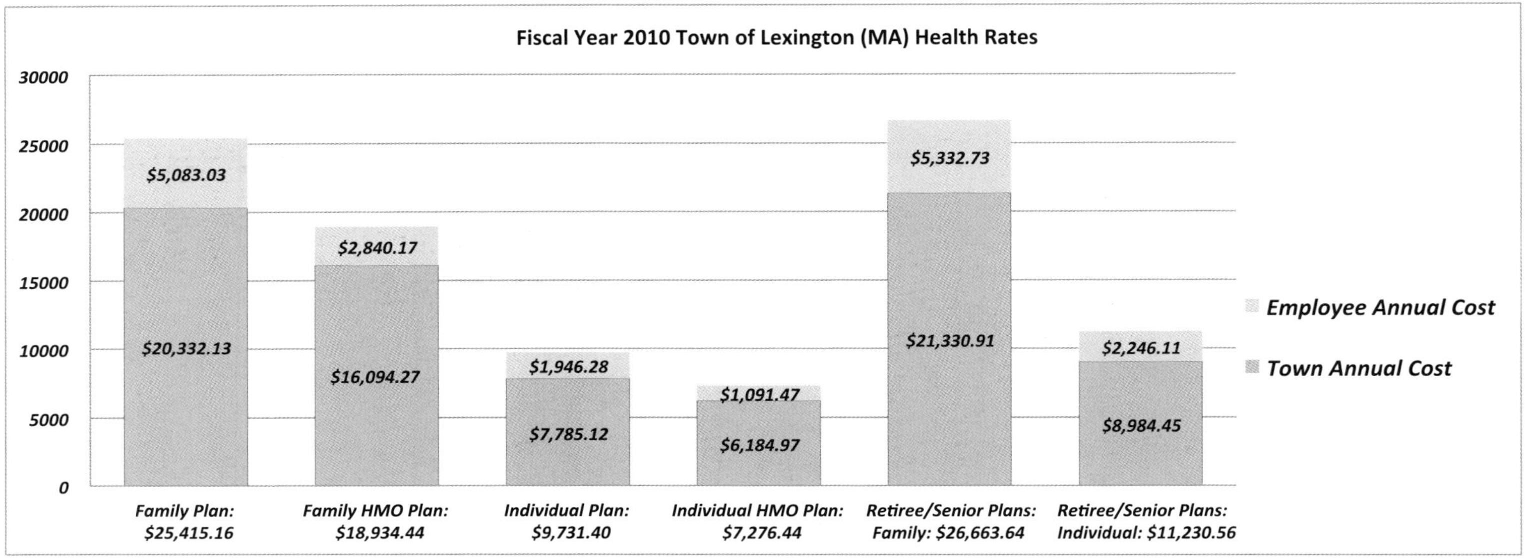

Fig. 5–3. Town of Lexington (Mass.) health/dental rates

In the public sector, health insurance is often a package deal for the municipality; thus, as a general rule, bargaining over a new health insurance plan cannot be based solely on the local union's wants and needs. Because of the impact on many other locals, significant plan changes may involve the municipality's bargaining with all of the unions, which must unanimously agree in order to make plan changes. Today, many municipalities are bargaining over lowering the percentage that the municipality pays for health insurance. This affects employees on two fronts: first, they will be paying an increase in the percentage of their insurance; second, each year they will be faced with an overall increase in the cost of the health insurance. Hence, it is important to consider how bargaining over individual items will impact other unions and work groups.

As will be discuss later (in chapter 9), the establishment of a wellness program and related health/safety incentives should be considered in the costing of the contract. Front-end money spent on a wellness program, by encouraging employees to make healthier lifestyle choices, is expected to lower medical costs down the road. The cost and saving ratio may differ from the first year to the second year. The return on this investment may be months or even years away.

Another cost impact that is often overlooked is when a wage item is provided now and a second wage increase is promised in the second year. For example, a 2% wage increase now and a 2% wage increase in the second year does not represent a 4% wage increase in true dollars. Keep in mind that the second year's 2% is based on the pay increase already given at the start of the contract, representing a wage increase of 4.04% from a true economic standpoint, not 4%, a common mistake. In a smaller department, the mistake of 0.04% for personnel may be absorbed without any concerns; however, in a fire department with several hundred employees, this 0.04% can result in significant budget issues.

Public sector contract costing is different than the private sector. In the private sector, it is about corporate profits and loss. The profit and loss factors may make or break a deal. Today, management and labor unions may use laptop computers with mathematical modeling and computer simulation to determine projected costs in a matter of moments while sitting at the bargaining table. Albeit quicker, poor data entry can result in the same costly mistakes, so it is best to have multiple levels of checks and balances in place when it comes to costing a labor contract. Very few people have the knowledge, skills, and ability to accurately predict the true cost of a newly agreed-on contract.

Contract language

The results of the hard work of bargaining will be reduced to writing, which will be placed into a document that is called a collective bargaining agreement (fig. 5–4). A poorly written contract, however, can undo all the hard work that labor and management put into establishing a tentative contract, leaving both parties frustrated over what they thought they agreed to at the bargaining table as compared to what is in the written contract. Mohamad and Madon have noted that "effective interpretation and understanding of the contractual obligations will be much dependent on the clarity of the document in translating the needs and requirements implied as in the agreement."[8] Poorly articulated contract language will result in many grievances, as well as in higher costs and time answering and defending grievances. In drafting contract language, keep in mind that the role of the arbitrator is to interpret the contract language, reflecting the spirit and intent of the article.

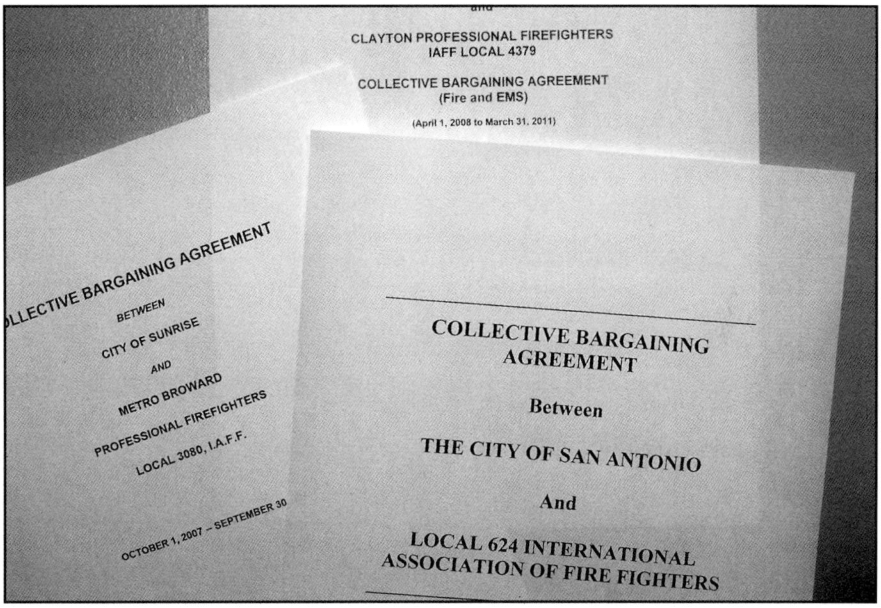

Fig. 5–4. A contract in writing. The collective bargaining agreement is the written, legally enforceable document containing all of the contractual langue that labor and management have agreed on for a specific period of time. The contract can range from a couple to more than a hundred pages in length.

As each party agrees to a given article at the bargaining table, the language that will be used in the given article should be crafted at that point in time, to ensure that the spirit and intent of the article is preserved. It is not recommended to wait, because leaving the contract language to the end of the bargaining session may result in difficulties and, possibly, a protracted process. Allowing each side to tentatively agree to the article language lets each side bring the article language back to the key players to ensure that it is workable and that nothing is missing; the two parties may mutually agree to tentative contract language and it may look bulletproof, only to find that the rank and file have questions or that the human resource department is unsure about the language. It is far better to clear up any clarity issues at this point. While the process of review may be challenging and time consuming, in return there will be far less time spent on grievances over the contract language.[9]

Once the newly drafted contract article language has been reviewed by different levels within the organization, it will often yield a clearly defined article. As the old saying goes, what makes sense at the table may not make sense on the floor.

Importantly, the written contract will be carried by some employees every day in their briefcases and may even be referred to multiple times a day. Thus, the terms and conditions contained in the contract should be written in a way that the average rank-and-file member will understand; a collective bargaining contract should not be written in such a way that only an attorney can interpret the language. Note, however, that labor and management may agree to ambiguous language, knowing that it may result in a grievance that could be settled by a neutral third party.[10] The bargaining team may also agree to ambiguous language with the intent of clarifying the language in a future contract negotiating session. In addition, as the bargaining team cannot foresee every possible work issue/event, changes in environment and external issues may change or obscure the meaning of contract language.

Three general categories of contract language ambiguities exist:

- Intentional
- Inadvertent
- Environmental/external

In the event that a grievance is filed over ambiguous language, the arbitrator will generally take into consideration the entire contract and not isolate the one article that is being called ambiguous. In some cases, other sections of the contract may clear up the ambiguity. In addition, if one party's interpretation is reasonable and the other's interpretation is unreasonable, then the arbitrator will side with the reasonable party. The bargaining team's words should have a common, everyday meaning understood by others within the work environment, rather than a specialized or extraordinary meaning. If words or phrases will be considered specialized or extraordinary, then the intent should be explicitly noted in the contract.[11]

Student Activity

- List guidelines that labor and management can use to avoid ambiguous contract language.
- Identify the best people within the organization who can aid in eliminating ambiguous language and why.
- Provide one example of using intentional ambiguous language and how it will help and/or hinder labor and management.

Evergreen clause. In the field of labor relations, the term *evergreen clause* is used to mean that the terms and conditions in the contract will be left in effect until the next contract is ratified, even after the expiration date of the current contract. Many labor unions will place this clause in the collective bargaining agreement to prevent management from changing anything in the contract, even after it has expired; any changes to the expired contract would have to be negotiated before any changes could take place. In other words, the contract is frozen in time, and no adjustments can be made by labor or management. This means no cost of living adjustments or increase/decrease in insurance coverage/cost unless first negotiated by labor and management. For example, in the event that the negotiation process is ongoing but the current contract has expired, the current contract would continue in force and effect until labor and management have ratified a new contract. In the case of the *Trustees of Sheet Metal Local No. 10 v. Genz-Ryan Plumbing and Heating Co.*, Civ. No. 08-3752 (July 2009), the federal court in Minnesota ruled that the employer is bound by the evergreen clause in the contract.

In October 2010, the Massachusetts Supreme Judicial Court (MSJC) issued a significant ruling that will have a profound impact on public sector bargaining in the state. The MSJC struck down the validity of the evergreen clause. A large majority of public sector collective bargaining contracts contain such language. In *Boston Housing Authority v. National Conference of Firemen and Oilers, Local 3*, SJC-10569, 458 Mass. 155 (October 22, 2010), the MSJC held that the contract language, extending the life of the contract until each of the parties reached a mutual agreement for a new contract, violates the Massachusetts General Law, as stated in Chapter 150E § 7(a): "Any collective bargaining agreement reached between the employer and the exclusive representative shall not exceed a term of three years. The agreement shall be reduced to writing, executed by the parties, and a copy of such agreement shall be filed with the commission and with the house and senate committees on ways and means forthwith by the employer."[12] Having an evergreen clause in the contract would violate the three-year limit on collective bargaining agreements established by the state legislature. For the past 25-plus years, public sector collective bargaining agreements have contained evergreen clauses, which were often approved by the Division of Labor Relations for the Commonwealth of Massachusetts. In this single ruling, the MSJC has significantly changed the landscape of public sector bargaining in Massachusetts, and the next chapter of public sector labor relations for Massachusetts is about to be written. Public sector unions and attorneys have been meeting round-the-clock to determine the best course of action and, in the immediate term, to protect those contracts that are currently expired under the evergreen clause. The next step is to file new legislation to remove the three-year limit from the law.

Negotiation Impasse

A bargaining impasse is when the two parties agree that there is no further benefit of bargaining, because no agreement is possible despite exhaustive good-faith bargaining. While each state may or may not define what requirements must be met for a bargaining impasse to exist, the NLRB has established the following definition for an impasse: "after good-faith negotiations have exhausted the prospects of concluding an agreement."[13]

Many states have adopted a no-strike section for firefighters, so going on strike is not an option during an impasse, unlike in the private sector. A number of states (Alaska, Connecticut, Delaware, Hawaii, Illinois, Iowa, Maine, Massachusetts, Michigan, Montana, Nebraska, Nevada, New Jersey, New York, Ohio, Oregon, Pennsylvania, Rhode Island, Vermont, Washington, Wisconsin, and Wyoming) have adopted some type of interest arbitration process for a bargaining impasse for firefighters and police officers and, in some cases, for other public employee groups. Generally, during the first bargaining session, the two parties will establish the ground rules and, most importantly, the process to be used in the event of an impasse. The current collective bargaining agreement may detail the steps to be taken in the event of a bargaining impasse, even if the current contract has expired. If this process is not determined in advance, the impasse will be further complicated by the two sides attempting to decide who may best settle the impasse.[14]

Generally, when an impasse has been reached, there are a couple of options that allow for a neutral third party's involvement. Public sector bargaining teams may be legislatively required to use one or all of the following options in the event of a bargaining impasse:

- Mediation
- Fact-finding
- Interest arbitration

Mediation

In the event that a bargaining impasse is reached, one option is to seek relief through mediation, wherein a neutral third party assists the two parties in formulating an agreement. There are 35 states that mandate mediation for public sector bargaining.[15] The mediator does not hold a decision-making position; rather, the mediator is empowered only to assist the two parties in reaching an agreement, as the two parties retain the right to either settle the impasse or continue with the impasse. While the mediator may suggest settlement avenues highlight common interests between the two parties, the final decision remains in the hands of the two parties themselves.

Mediation sessions are held behind closed doors, and each side is bound by strict confidentiality agreements. What is discussed or used during the mediation session cannot generally be used in later bargaining. Several states have laws pertaining to the discussions taking place during a mediation session, to ensure that they remain confidential and are not used in any other proceedings.

The mediator's goal is to bring a fresh set of eyes to the bargaining impasse and attempt to identify alternative settlement articles that may have been overlooked by the two parties. The key to this process is opening the lines of communication between the two bargaining parties. Generally, the mediator may meet a couple of times with the two parties individually to gain a sense of the parties' bargaining range and to find out how far off the two parties are from a settlement range. Additionally, the two parties undertake a cooling-off period once they decide that they reached an impasse. The mediation session may also be held in a neutral location so that neither party perceives that it has the upper hand. As Professor Edward Brunet and Professor Charles Craver have clarified, "what distinguishes mediation is the mediator will help the parties to resolve their disputes by aiding their ability to identify common interests and common ground for agreement. Although this process of identifying common goals and common areas of interest may seem like compromise, in fact it should be seen as each adversary realizing that it has a selfish, purpose motive to end the dispute."[16]

The mediation process can go a long way toward rebuilding the damaged relationship between labor and management, enhancing the bargaining process, and facilitating creative options for a settlement. If mediation is going to work, then each party must enter the process willing to make it work. If one or both parties enter the process with the intent of holding to their current stance regardless, then mediation will not work.

Fact-finding and interest arbitration

Fact-finding and arbitration have many strikingly similarities, yet they are two different options for resolving a bargaining impasse. Fact-finding is mandated in 31 states for public sector bargaining, in addition to 21 states offering interest arbitration for firefighters. Fact-finding and interest arbitration each involve the use of a neutral third party. Fact-finding is most often used in the public sector and is distinguished from arbitration in that the two parties are not obligated to accept the fact finder's ruling; in other words, fact-finding is not binding.

Interest arbitration, by contrast, is binding, with the arbitrator's decision being final. Whereas mediation and fact-finding involve compromise by the two parties, arbitration involves adjudication. Arbitration parallels the court system, where the final decision is offered by the judge (arbitrator), and, like the legal system, arbitration has a number of standards for operating procedures. However, the standards used in arbitration are far more flexible than those in our court system.[17] One large difference between arbitration and the legal system is the time involved in obtaining a ruling; arbitration is a much quicker and often less costly process. The procedures that labor and management have agreed on, the arbitrations organization, and the laws that have an impact on arbitration will all control the level of power that the arbitrator retains in the process. Hence, the power of the arbitrator depends on several factors.

Fact-finding and interest arbitration each follow a detailed process in preparing a settlement to a bargaining impasse. From a political standpoint, management can enjoy a certain freedom, since the arbitrator's award is handed down by a neutral third party; therefore, the political finger-pointing can be directed at the arbitrator instead of the administrator. Taxpayer organizations tend not to favor the use of arbitration because the neutral third party is not accountable to the taxpayers and has no vested interest in providing the best possible outcome. In line with this thinking, some believe that the law should not require arbitration in public sector disputes, which is called *compulsory arbitration*.

Importantly, the process of collective bargaining can be broken down into stages. In some cases, bargaining will progress through each stage, while in other cases, an impasse leads directly to a final stage, binding arbitration. The stages comprise a process that the two parties can use to settle an impasse. The stages of settling an impasse often reflect individual decisions between labor and management and will not be the same even for all the bargaining units covered within a municipality.

Stages of Settlement

1. Two parties meet for the purpose of collective bargaining, and agreement is reached. This is the early stage of settlement.
2. Impasse is reached: Two parties agree to use mediation or fact-finding with the assistance of a neutral third party (final decision remains with the two parties). This is the intermediate stage of settlement.
3. Arbitration: The two parties agree to go to arbitration to have a neutral third party decide the case. The two parties are compelled to accept the arbitrator's ruling as final and binding. This is the final stage of settlement.[18]

Final Thoughts

The process of preparing for and conducting labor negotiations is extensive. The intent of this chapter has been to expose the reader to the issue of negotiations and negotiation impasse. Because many readers will not be directly involved in the actual bargaining process, this survey has been relatively brief. It is recommended that a student who will be faced with the possibility of negotiating a collective bargaining agreement seek additional education and training on the topic. However, it is equally important to understand the labor relations process and how the bargaining process is interconnected.

No matter how long or how well labor and management work at the bargaining table, contractual language must reflect the spirit and intent of the two parties. This is not always an easy task to articulate in the contract. In some cases, a contract may contain ambiguous language that was incorporated either on purpose or inadvertently. There are steps that labor and management can take to prevent or minimize ambiguous language in the collective bargaining agreement. Having different levels of management, rank and file, human resource management, fiscal officers, and other stakeholders read and comment on ambiguous language will make for a better collective bargaining agreement, decreased grievances, and ultimately a better working environment.

Labor and management will often go through a detailed process well in advance of the bargaining process to determine areas of concern for the next negotiation session. The process may involve seeking input from the stakeholders, topic of most grievances, and inside/outside factors. Often, it is best if the two sides can narrow down the items to be discussed and bargained over. Because the bargaining process involves humans, emotions and ego can play a role in the process and can impede progress. Sometimes, labor and management may not be able to reach an agreement despite the efforts of bargaining in good faith, which is called an impasse.

It is during the bargaining process that labor and management will progress through bargaining stages. The two parties may advance through each stage, while others may reach an impasse and go directly to arbitration, avoiding the use of mediation and fact-finding. In mediation and fact-finding, an outside neutral person aids the two sides in brokering a settlement, with the ultimate decision being left in the hands of the two parties. Interest arbitration is when the arbitrator issues the decision and it is final and binding.

The prior chapter discussed the items that will bring labor and management to the bargaining table and legal requirements. In this chapter, how things are conducted during the negotiation process has been detailed. The next chapter will identify the critical elements to administering a collective bargaining agreement.

Key Terms

- 80/20 rule
- Bargaining impasse
- Bargaining team
- Chief negotiator
- Compulsory arbitration
- Contract costing
- Contract language
- Drop-dead-offer
- Evergreen clause
- Fact-finding
- Fractional bargaining
- Good-faith bargaining
- Interest arbitration
- Mediation
- Negotiating skills
- No-cost items
- Open-meeting law
- Time not worked

Exploring Issues on the Web

American Arbitration Association: http://www.adr.org/

The Association for Conflict Resolution (ACR): http://www.acrnet.org/Default.aspx (Professional organization enhancing the practice and public understanding of conflict resolution)

Fact-Finding: Practical Guidelines and Steps for Getting Started: http://www.adr.org/sp.asp?id=35765

New York State Dispute Resolution Association (NYSDRA): http://www.nysdra.org/Default.aspx (Not-for-profit association serving the needs of the dispute resolution field)

Southern California Mediation Association: http://www.scmediation.org/ (Founded in 1989 in response to the growing interest in mediation as a profession and as an effective adjunct or alternative problem-solving method)

References

1. Herman, E. E. 1998. *Collective Bargaining and Labor Relations*. 4th ed. Upper Saddle River, N.J.: Prentice Hall. P. 153.
2. Chandler, T., and R. Gely. 1994. Union and management organizational structures for bargaining in the public sector. In *Handbook of Public Sector Labor Relations*, ed. Jack Rabin. New York: Marcel Dekker. P. 37.
3. Thompson, L. 2005. *The Mind and Heart of the Negotiator*. 3rd ed. Upper Saddle River, N.J.: Pearson. P. 141.
4. Malhotra, D., and M. Bazerman. 2008. Psychological influence in negotiation: An introduction long overdue. *Journal of Management*. http://jom.sagepub.com/cgi/content/abstract/34/3/509 (accessed January 5, 2010).
5. Reitz, M. 2007. Public access to public sector collective bargaining. http://www.effwa.org/main/article.php?article_id=2107 (accessed September 8, 2010).
6. Holley, W., and K. Jennings, and R. Wolters. 2005. *The Labor Relations Process*, 8th ed. Mason, Ohio: South-Western. P. 254.
7. Granof, M. 1973. How to Cost Your Labor Contract. Washington, D.C.: Bureau of National Affairs. Pp. 2–3.
8. Mohamad, I., and Z. Madon. 2006. Understanding contract documentation. Proceedings of the 6th Asia-Pacific Structural Engineering and Contraction Conference. Kuala Lumpur, Malaysia. http://eprints.utm.my/515/1/MohamadIbrahimMohamad2006_Understanding_contract_documentation.pdf (accessed September 2, 2010).
9. Herman 1998, p. 177.
10. Cummings, F., and J. Hart. 2001. Labor contract clauses from A to Z. http://files.ali-aba.org/thumbs/datastorage/lacidoirep/articles/PL_CUMMINGS-01-1436_thumb.pdf (accessed September 1, 2010).
11. LaRocco, J. 2004. Ambiguities in labor contracts: Where do they come from? *Dispute Resolution Journal*. 59 (1): 38–41.
12. Commonwealth of Massachusetts. 2010. Mass. Gen. Law, Chapter 150E § 7(a). http://www.malegislature.gov/Laws/GeneralLaws/PartI/TitleXXI/Chapter150e/Section7 (accessed October, 28, 2010).
13. *Taft Broad. Co.* 163 N.L.R.B. 475, 64 LRRM 1386 (1967).

14. Holley et al. 2005, p. 593.

15. Ibid., p. 593.

16. Brunet, E., and C. Craver. 1997. *Alternative Dispute Resolution: The Advocate's Perspective*. Charlottesville, Va.: Michie Law Publishers. P. 189.

17. Summers, R., and R. Hillman. 1997. *Contract and Related Obligation*. 3rd ed. St. Paul, Minn.: West Publishing.

18. Elkouri, F., and E. Elkouri. 1999. *How Arbitration Works*. 5th ed. Washington, D.C.: The Bureau of National Affairs.

CONTRACT ADMINISTRATION

6

Learning Objectives

After reading this chapter, you should be able to:

- Identify the key points to contract-administering duties and responsibilities for a labor contract.
- Describe the alternative dispute resolution programs used in the public sector.
- Explain the grievance process.
- Define how duty of fair representation applies to a union member and the union.
- Describe how past practice is used in the fire service.

Contract Administration

Chapters 4 and 5 addressed the process of bargaining and negotiation. This chapter addresses the issues that come into play after labor and management have signed a collective bargaining agreement, when the real challenge of day-to-day administration of the contract begins. The administration of the contract can be a more daunting task than the negotiation process.

Some state labor laws govern the articles that are subject to bargaining, and some go so far as to look at the bargaining process even after the contract is signed. Having a signed collective bargaining agreement does not relieve the employer or the union from the aspects of state law that govern collective bargaining. Once a contract has been executed, it is the responsibility of each party to uphold the terms and conditions as specified in the collective bargaining agreement. The actions and attitudes from labor and management can play a role in how the day-to-day contract administration will progress. Stalled negotiations or a contract impasse can bring a cacophony of verbal attacks from one side, which serve no purpose in public than to agitate the other party. This can easily carry over into the day-to-day administration of the contract. Contract negotiation and administration should be conducted behind closed doors in a businesslike manner. The goal of contract administration for each side should be to administer the terms and conditions effectively and with integrity, otherwise the labor relationship falters.

As previously discussed, collective bargaining agreements may either be meticulously detailed and lengthy or be only a few pages in length. Regardless of the size and extent of the contract, labor and management have an equal responsibility to administer the contract. During the process of administering the collective bargaining agreement, there will be times that labor and management will have a different opinion or interpretation of the terms and conditions specified in the contract. A majority of collective bargaining agreements specify the means for settling these differences. The following pages will detail the grievance and arbitration processes used by labor and management to settle contract interpretation differences. Importantly, the dispute resolution process may be based on a collective bargaining agreement; however, the very same principles can be applied to nonunion fire departments. Like negotiation sessions, arbitration, mediation, and grievances are held in private, and only after the fact are the end results made public. Many times, the two parties are attracted to the dispute resolution process because of its private nature. Regardless, employees should be offered a process to have disputes resolved in a professional, unbiased, and quick manner.

Labor unions have a long and rich history when it comes to grievances. For many labor unions, the only way to have a dispute resolved was to file a formal grievance. However, many fire departments today use different methods to resolve disputes before they reach the official grievance process. In the following pages, recommendations and solutions will be reviewed as to how fire departments can take steps to resolve a dispute in its early stages.

Student Activity

- Research your state public sector labor laws/regulations and explain how labor arbitration works for resolving disputes involving firefighters.
- What are the scope and limitation to your state labor laws/regulations pertaining to dispute resolutions for firefighters?
- Does your state require any alternative dispute resolutions requirements pertaining to firefighters?

Laws and regulations pertaining to public sector bargaining and arbitration vary widely across the United States. It would be impossible to explain each state's arbitration laws in the limited space we have in this text. In some cases, the state will grant the right for the municipality to enter into a collective bargaining agreement with the union, yet the state will not allow the agreement to be binding between the two parties. Because the terms and procedures explained in this chapter may or may not apply in your state, it is recommended that you research the laws and regulations pertaining to arbitration in your state. The intent here is to describe in broad terms the various methods used for resolving labor disputes. The department's personnel management program should include some form of a dispute resolution process.

Administering duties and responsibilities

It goes without saying that labor and management are jointly responsible for the administration of the collective bargaining agreement. In simple terms, it is management that will often take steps in administering the fire department (action), while it will be the union that files a grievance over a contract violation (reactive). The union will be the predominant side to file grievances in response to the decisions that management will take in the administration and operation of the fire department. The union can serve as checks and balances on the administration of the department within the terms and conditions outlined in the contract. Many states require municipalities and labor unions to reduce the terms and conditions of implement to a written document, called a collective bargaining agreement. In public sector bargaining, some states require the agreement to be in writing and hold it to be binding; other states require it to be in writing but do not hold it as binding; and, finally, still other states make it illegal for the municipality to enter into a collective bargaining agreement.

In the states that require a collective bargaining agreement to be reduced to writing, this is often referred to as a *duty to sign*. Public and private sector labor laws consider the signing of the collective bargaining agreement as the final stage in the bargaining process. In some cases, refusing to sign a written contract is grounds for an unfair labor practice. As a point of clarification, a collective bargaining agreement is not an employment contract; an employment contract is between the employer and employee. The collective bargaining agreement is between the employer and the labor union representing the employees. The collective bargaining agreement will define the role of management and the role of the labor union (as discussed earlier in this book). The contract may extend certain rights to the employee such as job security, agency fee, and so forth.

The contract will not be worth the paper it is written on if no one administers the contract. When labor and management do not take the necessary steps to administer the collective bargaining agreement, the contract serves no purpose to either party. If management or labor is allowed to do what it wants without the other party filing a complaint or filing a grievance, the action could turn into a past practice, as we will discuss later in this section. The lack of administration of a collective bargaining agreement may be the result of a cultural issue within the organization, interorganizational conflict, or hierarchical dysfunction between labor and management. Regardless of the reason for the lack of enforcement, it is recommended that each side take the steps to begin the administration of the contract. To move forward, changes within the organizational hierarchical structure of both labor and management may be needed. It may take time to change this culture, since it was not developed in a day or a week. Slow, methodical, and consistent steps toward administration of the contract will move each party in the correct direction.

The written terms and conditions contained in a collective bargaining contract serve as the cornerstone to the labor relations process and personnel management. The contract will spell out the terms and conditions of employment to ensure that the employee, management, and the union have a clear understanding of the details. The contract provides for a written framework that all parties can work from and within. In essence, the contract is, to some degree, controlling and/or limiting human behavior of those covered under the collective bargaining agreement.

The collective bargaining agreement is a legal contract between labor and management and is often enforceable by the state or the legal system. The collective bargaining agreement must have a term period; contracts cannot be signed and live forever, so there a starting point and an ending point must be specified. A majority of collective bargaining agreements have articles that will detail how the contract renewal will take place and the advance notice served on the party, called a *renew notice*.[1] The renew notice period allows each side time to begin the bargaining process before the contract expiration date. The goal of each party is to ratify a contract before the expiration date; however, this is not always possible. In some cases, labor contracts are allowed to expire, and the bargaining continues with very limited progress, falling short of ratification. As a result of the economic downturn of the past couple of years, many labor unions are working under expired contracts.

When the rank and file discover that a possible contract violation has occurred, the contract does not allow the members not to follow the orders. Only in the event of a safety issue could a firefighter not follow an order and not be formally disciplined; members not following management orders should expect to be brought up on charges of insubordination. The contract provides a written framework for labor and management to follow, and in the event of a contract violation, the contract should also identify the steps needed to resolve the contract violation. Firefighters do not have the right to stage a walkout in protest of an order issued by management; rather, the preferred action is to follow the orders issued by management and immediately report the incident to a union official.[2] In larger fire departments, the union will often have a union steward on each shift or in each station; in smaller fire departments, the union member may need to call a union officer at home because the union membership is not large enough to support having a union officer always on duty. It is the responsibility of the union to ensure that the members know how to always contact a union steward or union official.

The remedies for contract violation range from a cease-and-desist order up to and including lost wages and legal costs, depending on the contract violation and the number of members affected. For example, I had a grievance filed by the union, seeking the removal of a captain's exam for a promotion. In reviewing the contract prior to posting the promotion, I had found it to be silent on the issue, so I implemented the exam requirement, knowing full well that this was a change in working condition and subject to bargaining; nevertheless, I wanted to test the water to see if the union would allow this new requirement for promotions and to enhance the credibility of the promotional process. At the first step in the grievance process,

I denied the union's request to remove the exam, and only at the second step did I agree to remove the requirement, in front of the town manager. Although the union walked away thinking it had won, I knew in advance I did not have the legal standing to require the exam and that my boss, the town manager, had advance notice of the action. However, if the union had not said anything about the newly required exam, then I would have implemented it and required it on all promotions moving forward. Almost 10 years later, the union and the fire department still have not changed the promotional process. Promotions are based on strict seniority rights; hence, the most senior person is the next to be promoted. It is sad that, in this day and age, seniority is the key factor in obtaining a promotion, even though the most senior person is not necessarily the most qualified to perform the assigned duties.

Just because something is not defined or covered in a collective bargaining agreement does not allow it to be handled exactly as management wants. There are certain exceptions to this general rule. If the law in the state requires bargaining to impasse on a change such as the promotion, then labor and management could choose to open up this one article for midterm bargaining; in the event that the two sides could not reach an agreement and bargained in good faith, management could implement the change, once the impasse is reached. This is all subject to the laws that the two parties are bound to within the state; many collective bargaining agreements have an article called an *opener* that details how the agreement can be mutually opened, during the term of the contract, to bargain over an issue/item.

Grievance Procedure

Justice requires a fair tribunal, but not necessarily an "optimal" one.

—Justice Stephen Breyer[3]

The grievance process comprises a series of steps that are followed in an effort to settle a dispute between the union and management. The grievance process can also be used in a nonunion setting to settle disputes between employee and employer. A grievance process could be incorporated into the employee handbook or the fire department policies/regulations. Personnel management in a call and volunteer department should allow firefighters be afforded a process for filing and resolving a dispute. Volunteer and call fire departments may establish a grievance committee made up of members who are appointed and/or elected and represent a mix of labor and management. Importantly, nonunion members (agency fee payers and/or free riders) who are covered under the bargaining unit (collective bargaining agreement) also have the right to the grievance process as outlined in the contract. In the public sector, it is common to have arbitration as the final step in the grievance process or another alternative dispute resolution process.

The grievance and arbitration process can be implemented in various settings when two parties have a disagreement. The grievance and arbitration processes provide the means for labor and management to settle disputes over the terms and conditions outlined in the contract and bargaining impasses (fig. 6–1). Each individual collective bargaining contract will establish the steps to be taken to resolve disputes. While it is common for a contract to address only impasse resolution, other labor contracts may be silent on how a contract impasse will be resolved, leaving it to the state labor agency/board's discretion.

A majority of labor contracts have grievance procedures for resolving contract violations. When discussing dispute resolution, each party should consult the collective bargaining agreement to ensure that the agreed-on terms are followed. Contracts dispute articles are not always the same. The grievance process, regardless of how many steps are included, will provide an opportunity for labor to present the interests of its membership and for management to present their side of the argument. In simple terms, the grievance process can serve as a pressure-relief value for the labor-management relationship.[4]

While grievances are often viewed as negative actions, this is not always the case. Labor and management should view the grievance process as a means for two parties to settle a dispute. In some cases, labor and management will use the grievance process to settle a dispute that could not be finalized at the bargaining table. Often, the grievance award will provide further guidance as labor and management move forward. Unresolved disputes, poorly designed contract language, lack of contract administration and department policies that have not been updated to reflect the changed organization can all lead to failure of a labor relationship and of personnel management. Labor and management must be willing to accept and honor awards issued by arbitrators, legal challenges issued by the courts, and policy changes to better prepare the organization for the changing world.

A number of public sector employees are covered under each state's civil service regulations. The goal of civil service was to put an end to party patronage from management in the public sector. Civil service regulations address issues surrounding appointment, classification, promotion, and discharge. It has sometimes been argued that civil service regulations and the grievance process outlined in collective bargaining agreements are redundant, allowing employees a second opinion in the due-process system. For instance, depending on the state and the collective bargaining agreement, a firefighter could lose a grievance on a discipline issue yet win at a civil service hearing, resulting in the discipline charge's being reduced or overruled. Although it may appear that, in some cases, firefighters have two options to seek relief over the same issue, the civil service program does not address many aspects of the labor relations process. The civil service process was not established to deal with day-to-day operational issues and concerns arising in labor-management relationships, whereas the grievance process was crafted to address disputes arising between labor and management.[5] It would be unfair to suggest that one or the other should or could be eliminated; however, it is fair to suggest that the redundancy built into the process tends to favor the employee.

Grievance Form for the Boomtown Fire Rescue Department
Boomtown Local 444 and the City of Boomtown

Name of Grievant (Please Print): **Job Title:** **Date of Hire:**	**Work Phone:** **Home Phone:**
Home Mailing Address: **Street:** **City:** **State:** **Zip:**	**Work Mailing Address:** **Street:** **City:** **State:** **Zip:**

Date, time and place of event leading to grievance:

Detailed description of grievance including names of other persons involve:

Explains the section(s) of the collective bargaining agreement that have been violated and the number of people who have been grieved by this action:

Detailed proposed solution to grievance:

Grievant: File a copy of this form with your immediate supervisor and retain a copy for filing at the next step or steps. If you do not receive a response within 10 working days or disagree with the action taken, you may file a copy of the grievance at the next step.

Step	Grievance Filed With *(Please Print Name)*	Date	Grievant's Signature	Date
1				
2				
3				
4				

It is recommended that the Grievant immediately contact a union representative to discuss the events and how to fill out this form.

Fig. 6–1. The grievance process. A formal process is used by labor and management to settle disputes arising from the collective bargaining agreement. Grievance procedures are considered unique to each contract.

While efforts should be made at the state level to eliminate or streamline the redundancy of the dispute process used by public sector employees, any such effort should expect to meet with prompt action from the public sector unions; the labor unions will pressure elected officials not to change the state civil service laws or the collective bargaining laws addressing labor disputes. From the point of view of the general taxpayer, it appears that the employee who is disciplined is allowed the protection of both the contract grievance process and the civil service protection hearings, double protection from discipline that is not offered to private sector employees. In each case, public and private sector, the dispute can be taken to the legal system to seek relief.

In most cases, the grievance process is initiated by the individual member or the union. A person who files a grievance is referred to as the grievant. The union can also file a grievance on behalf of its members. In the case in which the union files on behalf of its members, the stigma of one person's signing his/her name to the grievance and the possibility of retaliatory action to the person signing the grievance are avoided.

The grievant(s) may feel that management has violated the terms of the collective bargaining agreement and is seeking some type or form of relief, which may range from a cease and desist to being made whole again (the grievant must be made whole for all of the damages resulting from the contract violation). It is important to identify not only the violation but also what relief is being sought. The grievance process can begin with a simple oral complaint to a supervisor about the violation, or it may be in the form of a written grievance. Many public safety unions have a standard grievance form that must be filled out and signed by the member(s) or the union representative. The goal of the union representative should be to weed out the irrelevant and picayune grievances. Having a person sign a written grievance form may prevent frivolous grievances from being filed. A number of grievances are often settled at the first or second step of the process because of a misunderstanding or an oversight between the two parties. A significant factor in the early stage of settlement is how willing management and labor are to sit down and talk over the issue and their willingness of compromise. In review of past grievances from the union, if a majority of grievances are taken past the first couple of steps of the grievance process, a pattern may emerge—a single supervisor, a given article in the contract, or a couple of union members filing the majority of grievances. Labor and management should continuously monitor the reasons for grievances and the reasons for settlement. Monitoring these areas may reveal problem areas of supervision and ambiguous contract articles that could be used to correct these issues and may need to be brought to the bargaining table. Monitoring the dispute process may help inappropriate tasks that have been assigned or ordered, communication failures within the organization and between supervisors and employees, questionable personnel practices, or even a poorly designed process.

In some cases, depending on the significance of the grievance, labor and management may allow the union to bypass the first couple of steps of the process outlined in the contract. If the city manager changed the health insurance for the city without bargaining and without notice, the union may ask that the grievance be immediately taken to the final step, meeting with the city manager or going directly to arbitration. In this example, taking the grievance to the members' immediate supervisor, which is the first step, will not change anything, given that the supervisor does not have the authority to change the health insurance issue; further, since the next step may be the fire chief, who also does not have the authority to change the health insurance, the union might ask management to streamline the grievance process by taking the issue directly to the person who has the control and authority to redress the issue. In very rare cases, labor and management may mutually agree to take a grievance directly to arbitration and only use the last step of the process. In the event that the grievance process does not allow arbitration for dispute resolution, the two sides can mutually agree to take an issue directly to arbitration, but this would be limited to very rare instances. Some collective bargaining agreements may have mediation as the final step in the grievance process. The dispute resolution process selected by labor and management should be conducive to the relationship between the two parties.

The human resource department for the municipality should be involved in the grievance process. The human resource department may be able to provide training to managers and supervisors on resolving grievances. Human resource management should provide training and education on any contractual changes to all supervisors and managers after a new contract has been ratified or labor and management have agreed a change in the working environment. Human resources may also advocate for a satisfactory solution to the grievance that each side can live with. Incorporating human resources into the process can serve as a means to prevent grievances and facilitate their resolution.

No standard grievance process exists that is used in all cases; rather, grievance procedures are crafted by labor and management for inclusion in the collective bargaining agreement. Grievance procedures are considered unique to each contract. A contract may have only one step before arbitration, while another contract may have five steps before arbitration. The contract will often outline the formal steps and the time requirements for each step, for submission and decision. Some contracts require that the first step in the grievance process be written and presented to management; others require only a verbal presentation to the first-line supervisor about the dispute. In some cases, the collective bargaining agreement will detail the time frame for submitting a grievance. Not all contracts will have a time limit for filing. The union's collective bargaining agreement that I worked with as a fire chief did not contain a time frame, so the union could file a grievance over an alleged contract violation a year ago, for example; without a defined time frame, the door is wide open for the union to grieve about whatever it wants whenever it wants. This practice should be avoided, because the union may stockpile alleged grievances and, if it does not agree with the fire chief, threaten to file multiple grievances in an effort to show the municipal leaders how bad things are and how unwilling the fire chief is to resolve these issues. Fortunately, many municipal officials can see through such lame tactics to discredit management.

Regrettably, when the grievance process is abused and not used for its intended purposive, it can damage the labor relationship. When the grievance process is exploited in an attempt to discredit the management of the department or to place political pressure on municipal officials, it takes away from the credibility of the arbitration process. The goal should be to resolve a dispute as soon as possible, as allowing issues to sit serves no purpose from an organizational development standpoint. People want an answer to their dispute and to move onward. When labor and management have a defined time line for submitting grievances, it promotes a healthier work environment because disputes are addressed quickly.

Town of Farmington (Conn.) and International Association of Firefighters, Local 3103

ARTICLE 11

GRIEVANCE PROCEDURE

SECTION 11.1: The purpose of the grievance procedure shall be to settle employee grievances at as low an administrative level as is possible and practicable, so as to insure efficiency and employee morale.

SECTION 11.2: A grievance for the purpose of this procedure shall be considered to be an employee complaint concerned with:

SECTION 11.2.1: Discharge, suspension, or other disciplinary action.

SECTION 11.2.2: Charge of favoritism or discrimination.

SECTION 11.2.3: Matters relating to the interpretation and application of the articles and sections of this Agreement.

SECTION 11.3: Any employee may use this grievance procedure with or without Union assistance. Should an employee process a grievance through one or more steps provided herein prior to seeking Union aid, the Union may process the grievance from the next succeeding step following that which the employee has utilized. No grievance settlement made as a result of an individually processed grievance shall contravene the provisions of this Agreement.

SECTION 11.3.1 STEP ONE.

Any employee who has a grievance shall reduce the grievance to writing and submit it within ten (10) working days of the date of the occurrence to the head of the department who shall use his/her best efforts to settle the dispute. The department head's decision shall be submitted in writing to the aggrieved employee and his/her representative, if represented, within ten (10) working days of receipt of the grievance.

SECTION 11.3.2 STEP TWO.

If no agreement is reached with the decision rendered by the head of the department, the employee or his/her representative shall submit the grievance in writing within ten (10) working days to the Town Manager. Within ten (10) working days after receiving such grievance, the Town Manager shall render his/her decision in writing to the aggrieved employee and his/her representative.

SECTION 11.3.3 STEP THREE.

If no agreement is reached with the decision rendered by the Town Manager, the Union (and the Union only) may submit the grievance to arbitration to the Connecticut State Board of Mediation and Arbitration ("Board") within ten (10) working days after receipt of the written decision of the Town Manager or if mediation services are used, within ten (10) working days after the conference with the State Mediator, and the decision rendered by the arbitrator or arbitrators shall be final and binding upon both parties.

SECTION 11.3.4 MEDIATION.

The mediation services of the board may be used after the second step above provided both parties mutually agree. A request for mediation shall be filed with the Board by the employee or his/her representative within ten (10) working days after receipt of the written decision of the Town Manager.

SECTION 11.3.5 MEETINGS.

If either of the parties related to the grievance process desire to meet for the purposes of oral discussion, a meeting shall be requested and if satisfactory to the other party, a meeting shall be promptly scheduled. Such procedure shall not extend the time limitations above.

SECTION 11.3.6 TIME EXTENSIONS.

Time extensions beyond those set forth in this grievance procedure may be arranged by mutual written agreement by both parties.

SECTION 11.3.7 RECORDING OF MINUTES OR TESTIMONY.

Either party shall have the right to employ at his/her own expense a public stenographer at any step of this procedure. The party who employs a stenographer must provide a copy of a transcript of the minutes or testimony recorded to other party.

SECTION 11.3.8 REPRESENTATION.

Either party shall have the right at his/her expense of choice of representation whenever representation is desired.

SECTION 11.4 Attached to this Agreement as Appendix is the agreed upon grievance form to be used by the Union and/or its members.

SECTION 11.5 In the event that the Town or designees thereof fail to answer a grievance within the time specified, the grievance may be processed to the next step.

Sources of grievances

The goal of the grievance process is to formalize the investigation and resolve differences between labor and management in the workplace. Most grievances reaching the arbitration stage are the result of discipline and discharge actions by the employer. Management may also file a grievance based on the actions of the union—for example, activities that amount to a strike by a large group of employees, as by calling in sick from work, called a *sick-out*.[6] A majority of grievances filed by the union are over the following issues:

- Violation of the terms outlined in the contract
- Disagreement over facts
- Disagreement over the meaning of an article in the contract
- Disagreement over the administration of the contract
- Disagreement over fairness
- Disagreement over due process

Student Activity

Evaluate your fire department's current dispute resolution process. Each student should answer the following questions. Then, in small groups, compare your answers with others and discuss. Each group should present a report summarizing their discussion.

1. What type of dispute resolution process does your department have?
2. Explain the process one would take in resolving a dispute.
3. Is the final step final and binding?
4. What is the appeals process?
5. What changes or improvements could be implemented in the current system?
6. How would you go about making changes/improvements to the system?
7. Provide an example of the proposed system that is being used in another department/agency.
8. What challenges would you face in implementing the proposed system?

Arbitration procedure

Arbitration is not considered a new approach to disputes. In fact, arbitration has been around since long before English common law. In the case of *McAmis v. Panhandle Pipe Line Co.*, 23 LA 570, 574 (Kansas City Court of Appeals 1954), the courts called arbitration "the oldest know method of settlement of disputes."[7] State and municipal government have incorporated the use of arbitration for individual grievances by workers who are not protected by a labor union or organization. Fire departments may be able to implement an arbitration process between the department and the employee. Consideration should be given to the use of arbitration in volunteer departments because disputes will arise over time. Departments may want to limit the types of disputes that can be submitted to arbitration (discipline and terminations) or leave the door wide open for any type of dispute between the department and the employee/volunteer. Regardless of the process selected, the fire department administration and the rank and file should work together in determining the best arbitration procedure for the department. When a fire department unilaterally implements a dispute resolution procedure, the balance of power between labor and management is often tilted in favor of management.

Historically, arbitration procedures date back to the passage of the Federal Arbitration Act in 1925, as the United States' most important legislation in support of arbitration law and policy. The Federal Arbitration Act was established on "written provision in any maritime transaction or a contract evidence a transaction involving commerce" and was initially not intended for labor contracts. The use of arbitration in labor-management disputes evolved over time, as will be illustrated later in this chapter through the examples of several landmark legal cases in support of the arbitration process to settle labor contract disputes. During World War II, the War Labor Board expanded the practice of compulsory arbitration in disputes.[8]

In the early years, one argument against the use of arbitration was that the use of a private person (a neutral third party) to decide governmental policy was unconstitutional, based on the delegation of government power (the sovereignty doctrine). However, in the past 40 years, the courts have issued rulings that have eroded the sovereignty doctrine. In fact, legislators and past U.S. presidents support some level of dispute resolution for the public sector.[9] In 1998, Congress passed the Alternative Dispute Resolution Act, requiring federal district courts to authorize, by local rule, "the use of alternative dispute resolution processes in all civil actions," demonstrating that alternative dispute resolutions are a viable option instead of seeking legal action. These and other legislative measures have allowed arbitration as a procedure to resolve disputes, enforcing arbitration awards.[10] Many states that have adopted grievance procedures looked to the Federal Arbitration Act for guidance in establishing public sector grievance procedures. Regardless of the procedures selected, it is critical that the collective bargaining agreement stipulate which process will be used in settling disputes. In some cases, the collective bargaining agreement may also define the rules the grievance procedure will follow. For example, labor and management may mutually agree that the arbitration rules will follow the Labor Arbitration Rules of the American Arbitration Association, a not-for-profit organization.

American Arbitration Association, Labor Arbitration Rules

A detailed list of the rules for labor arbitration can be viewed online at http://www.adr.org/sp.asp?id=32599.

Before getting into the types of arbitration, we must distinguish between *rights disputes* and *interest disputes*. Elkouri and Elkouri explain the distinction as follows: "Disputes as to rights are adjudicable under the laws or agreements on which the rights are based, and are readily adaptable to settlement by arbitration. Disputes as to interests, on the other hand, involve questions of policy which, for lack of predetermined standards, have not been generally regarded as justifiable or arbitrable."[11] Thus, an interest dispute involves settling the terms of a new collective agreement (fig. 6–2). Some states will define how public sector disputes arising from rights or interests will be determined. In the most general terms, rights disputes can often be resolved in the form of adjudication, arbitration, or mediation-arbitration. Interest disputes are often resolved around mediation, economic power (many public sectors unions cannot go on strike), or arbitration. The public sector will often use interest arbitration as an alternative to strikes to settle disputes because government cannot tolerate a public strike or the lockout of employees by the government.

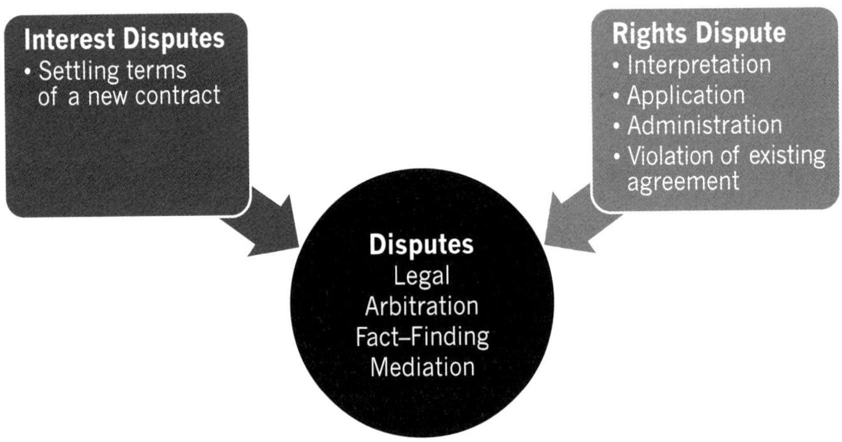

Fig. 6–2. Rights and interest disputes. Rights disputes fall under the legal or quasi-law standard. Interest disputes fall under the conflicting interests.

Several arbitration systems may be used in a labor arbitration proceeding. The arbitrator could be used in permanent or temporary bases in labor arbitration cases. The permanent arbitrator is often referred to as an *impartial chairperson* or *umpire* in the labor relations field. In the case of a permanent arbitrator, the arbitrator may serve for a set period of time, which is generally the length of the collective bargaining agreement or a period set by labor and management. This way, any issues will be submitted to the same person, who now has an established history and familiarity with all grievances pertaining to this contract, often resulting in a speedier hearing and decision. Moreover, the permanent arbitrator tends to issue awards that are consistent across the board. Labor and management have the ability to work with arbitrators, thereby gaining a sense of which cases will and will not hold merit with an arbitrator. The assignment of a permanent arbitrator is not without problems, as one party may find that the permanent arbitrator is providing awards that favor the other side, and it may be difficult at best to remove a permanent arbitrator. Permanent arbitrators who have worked with labor and management may also run the risk of knowing too much about the labor relationship and may be tempted to overstep the arbitrator's authority in offering awards.[12]

The second option is to have a panel of arbitrators to hear a case. An arbitrator panel is called a *tripartite panel*, in which the panel is made up of three arbitrators: one arbitrator is selected by management; another arbitrator is selected by labor; and the final arbitrator is a neutral arbitrator, often selected by labor and management. The tripartite panel can also serve as a permanent panel for a set period of time. Many of the benefits and risks of having a permanent tripartite panel are the same as those of having a permanent arbitrator. The permanent tripartite panel may be used at a state level in the public sector or in the private sectors when a company has a high rate of grievances and can support the use of a permanent panel. Call and volunteer fire departments may set up a form of dispute resolution that is modeled after the arbitration process and may have a tripartite panel that is made up of labor and management and a neutral third party.

The third option is the use of a temporary arbitrator to hear a single case. In this case, the arbitration rules or the collective bargaining agreement may define how single arbitrators will be selected. Generally, a list is provided by the arbitration association, or the state may offer a list of arbitrators. The state or the association has taken the time to qualify and select a large list of arbitrators, and it is up to labor and management to select one arbitrator from the list.

There are a couple ways in which labor and management can select a temporary arbitrator from such a list. One way is for labor and management to list their top three candidate arbitrators in private and then compare lists to see whether there is a name in common to both lists; alternatively, the strike method may be used, in which labor and management mutually agree on who are the least desirable arbitrators and strike their names from the list, until only one arbitrator is left on the list. In this method, it is important to start with an odd number of arbitrators, usually three or five, so that the parties get an equal number of strikes; also, from a tactical standpoint, the party who strikes second will have the final say, so it is best to offer first strike to the other side.[13] Hence, each arbitration case could result in a different arbitrator's issuing awards.

As noted in chapter 5, it is recommended that labor and management agree on how disputes will be resolved *before* a dispute arises. This will allow the best possible selection and minimize the influence of emotions, personalities, and actions on the selection process. Drafting a dispute resolution process before a dispute arises will allow labor and management to create a contract article that will withstand judicial review and will allow each party to determine the level of risk, cost savings, and efficiency in the dispute process.[14] The collective bargaining agreement may state that all disputes arising from the contract will be submitted to arbitration, or the contract may list only the items that will be submitted to arbitration.

Labor and management are often drawn to the arbitration process in an attempt to settle differences in a private setting, as compared to the public display of a judicial hearing. When disputes are presented in the court of law, the general public is exposed to all the details and evidence. Some might argue that presenting disputes in the public eye keeps the process transparent and open for scrutiny by the general public; however, arbitration hearings and evidence are kept private, and in many cases, the arbitrators' award is made public only after it has been awarded. Another favorable arbitration option is the streamlined process of seeking a ruling on a dispute, compared to the never-ending discovery process, complicated rules of evidence, and higher legal costs in seeking relief in the judicial system.

Most parties will agree to an ex parte hearing, which allows the hearing to be conducted in the event that the other party refuses to appear and has been served with due notice. Acceptance of this process will ensure the hearing is attended by labor and management. The following are considers when crafting an arbitration article in the contract or a *simple agree* to arbitration in a separate format:

- Dispute area to be covered by arbitration
- Will arbitration be final and binding?
- Will the case be presented to one arbitrator or a panel?
- What arbitration rules to be followed?
- Use of ex parte hearing
- Enforcement of the award
- Who is responsible for the cost of arbitration?

Most people think of arbitration as a labor-management dispute resolution process. Arbitration has direct roots to conflict resolution in the trade, commercial, and financial organizations. Not-for-profit organizations will often use the arbitration process to settle a dispute. Arbitration is also used commonly in the private sector: look at your credit card, phone, cell phone, and electric bills; you may be surprised to find that any dispute you have over your bill is subject to final and binding arbitration. Thus, the information contained in this chapter can help you professionally and personally. Arbitration is not just a union procedure, even though many people mistakenly think of it as such.

More than half the United States requires some form of arbitration as a means for settling labor management disputes. This law ranges from voluntary arbitration, when labor and management voluntarily agree to submit a dispute to arbitration. In this case, labor and management will decide beforehand how much weight the arbitrators' decision will hold. In some cases, the two parties agree that the arbitrators' decision is final and binding, while in other cases, the two parties will use the award strictly in an advisory manner. When an arbitrator makes a decision on a case, it is called an *award*.

As a result of the inability of most public safety employees to go on strike, labor and management needed another means for settling labor disputes. A majority of fire and law enforcement state labor laws require that contract disputes be submitted to compulsory binding arbitration. In this case, labor and management are required by law to present disputes to an arbitrator or a tripartite panel. In the event that firefighters and law enforcement unions could not put an end to a labor dispute, it would be the general public that would suffer the loss or hardship; therefore, the government has stepped in to ensure that the general welfare of the people is provided for. At the federal level, the Federal Service Impasse Panel can implement the terms of a dispute process under the Civil Service Reform Act of 1978. The postal service can be held to compulsory arbitration under the Postal Reorganization Act of 1970.[15] Interestingly, compulsory arbitration is just about nonexistent in the private sector, and the Taft-Hartley Act requires mediation only in disputes.

The argument of whether compulsory binding arbitration is good or bad for public sector labor relations will continue for years. On one hand, the argument against compulsory binding arbitration is that is does not allow labor and management to freely bargain over items, knowing that in the end a neutral third party will decide. If an arbitrator will be responsible for awarding such items as wages, benefits, and other terms and conditions of employment, public policy will tolerate only so much of a dictatorial process before governmental regulation will be implemented to limit the process of collective bargaining in the public sector. This places the true decision-making in the hands of someone who is not accountable to the taxpayers. Additionally, some argue that binding arbitration can breed labor relations discontent and disharmony.

A third common procedure is the use of final-offer arbitration in settling labor relations disputes. In this case, the arbitrator's award is based on a issue-by-issue basis, choosing between the final offers that each party has presented. The issue-by-issue form of final-offer arbitration is used in the states of Connecticut, Iowa, and Michigan.

In some cases, the two parties may agree to send several issues as a package to the arbitrator, and the arbitrator will again select one of the final offers (packages) presented by the two parties as the award. The package form of final-offer arbitration is used in Hawaii, Minnesota, Nevada, New Jersey, and Wisconsin. Each party will need to persuade the arbitrator or the panel that its final offer is more reasonable than the other party's. The arbitrator's award is based on the final offer, in contrast with the traditional process for arbitration of an impasse, where the two parties know that the arbitrator has the flexibility to impose his/her own award, so the desire is not always to present the best front. When labor and management have reached an impasse, they may use the final-offer or the last-best-offer arbitration processes because they tend to offer the greatest incentive for labor and management to settle differences or, at a minimum, provide the best offer and to settle disputes earlier than contracts with traditional arbitration.

Labor arbitration is used in disputes involving police officers and firefighters because of the inability to go on strike. Lacking any legal requirements, many municipalities seek labor arbitration, instead of court action, for the following reasons:

- Takes place behind closed doors in private.
- Flexible rules as compared to litigation.
- Selection of decision-maker generally up to labor and management.

- Reduced costs and faster decisions for dispute resolutions. Generally, the cost, time, and relaxed rules result in less money spent in arbitration than a lengthy legal case.
- Enforceable award and binding. The arbitrators' award is enforceable in court, and arbitrators' awards may be appealed in limited cases.

Some have argued that labor arbitration favors labor and results in less accountability at the local municipal level. Conversely, having a neutral third party decide the case without political pressure can also be a negative aspect for public sector unions. Interestingly, a majority of cases that are submitted to arbitration are settled before an award can be issued, because the two sides have had an opportunity to present information during the process and it allows them time to make adjustments. As figure 6–3, from the New Jersey Public Employment Relation Commission, shows, 45 interest arbitration cases were voluntarily settled in 2009 before the arbitrators could issue an award, as compared to 16 total cases in which the arbitrator issued an award. In addition, the average salary increase awarded in 1993 was 5.65%, as compared to a salary increase of 3.75% in 2009; as reported in the first six months of 2010, the average salary increase award was 2.43% (fig. 6–3). From these data, it would appear that it is best to go to arbitration, because the average increase is around 2.5%–3%, based on arbitrators' awards and the average salary increase of reported voluntary settlements.

Student Activity

Determine your individual state laws that pertain to fire department collective bargaining impasses and dispute resolution process. Because each state is different, it is important to identify the laws and regulation that govern public sector dispute resolutions.

1. Does your state have laws pertaining to collective bargaining? How will bargaining impasses be handled?
2. Once the department dispute resolution process has been exhausted, what steps does the state recommended or mandate? Arbitration, mediation, and/or fact-finding?
3. Does your state labor board oversee the arbitration process? What arbitration organization does your department use?

Table 6–1. New Jersey Public Employment Relations Commission salary increase analysis for interest arbitration from 1/1/1993 to 6/30/2010

1/1/1993–6/30/2010

Time Period	Total # of Awards Issued	Substantive Appeals Filed w/PERC	Average of Salary Increase All Awards	Number of Reported Voluntary Settlements	Average Salary Increase of Reported Vol. Settlements
1/1/201–6/30/2010	5[1]	2	2.43%	23[2]	2.80%
1/1/09–12/31/09	16	5	3.75%	45	3.60%
1/1/08–12/31/08	15	2	3.73%	60	3.92%
1/1/07–12/31/07	16	1	3.77%	46	3.97%
1/1/06–12/31/06	13	3	3.95%	55	4.09%
1/1/05–12/31/05	11	0	3.96%	54	3.94%
1/1/04–12/31/04	27	2	4.05%	55	3.91%
1/1/03–12/31/03	23	2	3.82%	40	4.01%
1/1/02–12/31/02	16	0	3.83%	45	4.05%
1/1/01–12/31/01	17	0	3.75%	35	3.91%
1/1/00–12/31/00	24	0	3.64%	60	3.87%
1/1/99–12/31/99	25	0	3.69%	45	3.71%
1/1/98–12/31/98	41	2	3.87%	42	3.77%
1/1/97–12/31/97	37	4	3.63%	62	3.95%
1/1/96–12/31/96	21	2	4.24%	35	4.19%
1/1/95–11/31/95	37	0[2]	4.52%	44	4.59%
1/1/94–12/31/94	35	0[2]	5.01%	56	4.98%
1/1/93–12/31/93	46	0[2]	5.65%	66	5.65%

[1]Does not include awards on appeal.
[2]Includes only settlements in impasses for which an arbitrator was assigned.

Arbitration and courts

The question is often raised about the finality of an arbitration award and the ability of judicial review. Elkouri and Elkouri have asserted that "errors of law or fact almost never justify vacating an arbitrator's decision. An arbitrator's award will not be overturned for errors of judgment on the law unless 'manifest disregard for the law' has occurred."[16] One feature that makes arbitration agreeable to each party is the reduced cost and length in receiving an award, as compared to seeking judicial relief. Many legal challenges have been filed over the issue of binding arbitration; however, the courts are reluctant to overturn or vacate an arbitration award. Nevertheless, in the cases that have been seeking judicial relief, the process can prolong the final award, further damaging the labor relationship between the two parties and leaving them in legal limbo. In addition, the added cost of seeking judicial relief can be as expensive as taking a dispute directly to court. If labor and management intend to have arbitration awards reviewed by the courts, then they should note this in the predispute agreement and/or in the collective bargaining agreement. In fact, the courts are the ones who will provide the legal foundation for turning arbitrators' awards into enforceable judgments.

In some cases, labor and management may detail the items that would allow for judicial review of an arbitrator's award. If granting judicial review, labor and management may empower the courts to modify an arbitrator's award. For example, the courts can correct a miscalculation of figures and/or assess whether the arbitrators ruled on subjects not submitted to them. As arbitrator John Wilkinson has stated that allowing modification of awards "has substantially added to the time and cost of arbitration while seldom changing the result, particularly since the U.S. Supreme Court has held, in *Hall Street Associates v. Mattel Inc.*, 128 S. Ct. 1396 (2008), that grounds for vacating an arbitration award under the Federal Arbitration Act are limited to unlikely occurrences such as 'evident partiality,' 'fraud,' 'corruption,' refusing to hear 'pertinent and material' evidence, and acts exceeding the powers of the arbitrators" (fig. 6–3).[17]

Fig. 6–3. Judicial review of arbitration awards. The courts are reluctant to overturn or vacate an arbitration award. The U.S. Supreme Court denied partitions for certiorari involving arbitration in several cases.

Many state labor laws also have sections limiting judicial review of arbitration awards. By allowing judicial review of any type of disagreement with the award, labor and management leave the door wide open for the losing party to appeal the award, and a full evidentiary appeal would be costly, timely, and harm the labor relationship.[18] In October 2009, the U.S. Supreme Court denied partitions for certiorari involving arbitration in the following cases: *Gain v. Trinity Health, No. 08-1446*, a decision by the 6th Circuit (551 F.3d 374, 2008); *Coffee Beanery, Ltd v. WW, L.L.C.*, a case decided by the 9th Circuit (300 Fed. Appx., 415, 2008); and *Comedy Club, Inc. v. Improv West Associate*, 6th Circuit (553 F.3d 1227, 2009). It is recommended that you review your individual state's laws pertaining to public sector labor arbitration, as well as the collective bargaining agreement for required arbitration steps that may be included in the agreement.

In the public sector, the power to enforce an arbitrator's award rests within the state and the court system. As discussed earlier, each state has determined the level of labor laws pertaining to the public sector, and these laws often dictate the enforcement of collectable bargaining agreements and arbitrators' awards. Some state labor boards/agencies have the authority to enforce arbitration awards; in other cases, the state requirement may be silent, leaving enforcement up to the courts. In 1957, in a landmark legal case, *Textile Workers v. Lincoln Mills*, the Supreme Court required arbitration when the two parties had perversely agreed to arbitration (meaning that the two sides had mutually agreed on arbitration as a means to settle any disputes). In this case, the union had entered into a collective bargaining agreement with the employer (Lincoln Mills of Alabama). The collective bargaining agreement stated that grievances would be handled through a specified procedure, with the last step calling for arbitration. However, when the union requested arbitration for a grievance, the employer refused. The union sued in federal court to demand that the employer fulfill the collective bargaining requirement calling for arbitration of the grievance. The basis for the decision was Section 301 of the Labor-Management Relation Act, which provided federal court review of violations stemming from collective bargaining agreements of industries affecting interstate commerce. The Lincoln Mills case is best known for requiring binding arbitration.

Lincoln Mills Case Review Online

To review the U.S. Supreme Court case Textile Workers v. Lincoln Mills, 353 U.S. 448, see http://supreme.justia.com/us/353/448/index.html

In 1960, the U.S. Supreme Court ruled on three cases dealing with arbitration, which would become known collectively as the "Steelworkers' Trilogy." Each of the three cases involved the United Steelworkers of America. The court addressed the relationship between arbitral and judicial processes. In a long list of legal rulings that place grievance arbitration as the submissive position, the U.S. Supreme Court, in the Steelworkers' Trilogy, added credibility to the grievance arbitration process.

In the first case, the court said that the courts should not look at the merits of the case; rather, it is the responsibility of the arbitrator, when the contract specifies binding arbitration. The decision whether an arbitration case has merit rests in the arbitrator's hands, not the courts, and it is up to the courts to determine whether the grievance is arbitral. In the second case, the Supreme Court said that unless a subject is explicitly excluded from arbitration in the contract, it leaves the door open for items to be subject to arbitration, when binding arbitration is specified. The U.S. Supreme Court elaborated,

> The function of the court is very limited when the parties have agreed to submit all questions of contract interpretation to the arbitrator. It is then confined to ascertaining whether the party seeking arbitration is making a claim which on its face is governed by the contract. Whether the moving party is right or wrong is a question of contract construction for the arbitrator. In these circumstances the moving party should not be deprived of the arbitrator's judgment, when it was his judgment and all that it connotes that was bargained for.[19]

The third case of the Steelworkers' Trilogy was much different in nature, involving the issue of whether courts can overrule an arbitrator's ruling. The Supreme Court said no, providing the arbitrator "draws its essence from the collective bargaining contract."[20] In this case, the court preserved the integrity of the arbitrator's award.

These legal rulings in the Lincoln Mills case and the Steelworkers' Trilogy set legal precedents for grievance arbitration in the field of labor relations. At the time, these cases made it clear to labor and management that arbitration awards are final and binding, with very limited review of the courts. Many states have used the standards established by the U.S. Supreme Court, in the Lincoln Mills and Trilogy cases, in crafting their public sector labor laws. Since the Steelworkers' Trilogy cases, the Supreme Court and federal appeals court have ruled in other major cases supporting the idea of arbitration, while leaving the door open for court reversal of an arbitrator's award.[21]

Steelworkers' Trilogy Case Review Online

For a review of the U.S. Supreme Court case *Steelworkers v. American Mfg. Co.*, 363 U.S. 564 (1960), see http://supreme.justia.com/us/363/564

For a review of the U.S. Supreme Court case *Steelworkers v. Warrior & Gulf Co.*, 363 U.S. 574 (1960), see http://supreme.justia.com/us/363/574

For a review of the U.S. Supreme Court case *Steelworkers v. Enterprise Car*, 363 U.S. 593 (1960), see http://supreme.justia.com/us/363/593

Online dispute resolution programs

Alternative dispute resolution has evolved into the online dispute resolution process, in which software programs have been developed to resolve labor-management disputes. The process of online dispute resolution, while nontraditional, provides another option for resolving disputes. The online dispute resolution process is an outgrowth of an e-commerce dispute program (fig. 6–4). One organization leading the march of technology to online dispute resolution is Federal Mediation and Conciliation Services, whose mediation division experimented with parties filing claims, submitting evidence, and tracking documents submitted from the other party. The process also allowed for online communications between the mediator and the parties.[22] This is an emerging technology in the field of labor relations, as well as mediation and arbitration.

Fig. 6–4. Online dispute resolution, an emerging technology in the field of labor relations

One interesting aspect of the online process is the minimizing of emotions that are evident in physical, face-to-face meetings. These emotions may be between the two parties or toward the mediator and/or arbitrator. The feeling of bias is also minimized by not having names to put to faces in the online process. Even if not conducted entirely online, the dispute resolution process could be a hybrid version, with each party submitting a majority of the material online and, if needed, having a physical hearing on location.

The evolution of online dispute resolution will be worth watching. At present, it does not yet appear to have surpassed the traditional mediation and arbitration process.

Fair Representation

A hot-button issue with any union is fair representation. As the exclusive bargaining representative of the workers (as outlined in local public sectors bargaining laws, the National Labor Relations Act, and the Railroad Labor Act), the union has an obligation, based on Supreme Court rulings, to represent all members covered under the bargaining unit fairly and without discrimination. The Supreme Court has held labor unions to the standard of fair representation, throughout the bargaining process and contract administration.[23]

The union will strive for the best interest for the majority, though it may at times conflict with the interests of the minority or of individual members. Hence, the labor union has been granted the exclusive bargaining interest of the group of employees, and individual employees cannot bargain directly with the employer. By bestowing this power the unions, the U.S. Supreme Court has created the duty-of-fair-representation doctrine. Some states have incorporated a level of fair representation into their public sector bargaining laws and have granted the state labor agency to investigate and enforce the duty of fair representation. The federal courts can award injunctive relief and damages. The goal is to ensure that a union represents all its members in a fair manner and without discrimination.

The root of fair representation can be traced to the country's history of racial discrimination. In the case of *Steele v. Louisville & Nashville Railroad Co.*, 323 U.S. 192 (1944), a black employee was subjected to blatant racial discrimination in the seniority system. Back then, the country did not have any laws preventing discrimination. Remember, there was no federal agency to investigate or enforce the Railway Labor Act of 1926. It was determined that the federal courts could hear the case and award damages and/or any other injunctive relief, to prevent any union from discriminating against a race or person.

The next most notable case of fair representation was Vaca v. Sipes, 386 U.S. 171 (1967). In this case, an employee named Owens had been out of work on disability. Owens presented the employer with a doctor's note stating that he could return to work; however, when the employer refused to allow Owen back to work, he filed a grievance with the union. On behalf of Owens, the union processed the grievance and assisted him in obtaining the needed medical records. Nevertheless, the employer denied the grievance and would not allow Owens back to work. When Owens requested that the grievance be taken to arbitration, the union denied his request, feeling that the case was not strong enough because the collective bargaining agreement did not allow individuals to bring a case to arbitration; only the union could file for arbitration. Owens filed a claim in state court against the union for refusing to take his grievance to arbitration, and the state court agreed that Owens had a valid grievance and awarded damages (unfortunately, Owens had died before the ruling was handed down).

Later, the Supreme Court ruled that the state court did not apply the rule in the case of *Vaca v. Sipes*. The Supreme Court said that the union will only breach the fair representation if the conduct is "arbitrary, discriminatory or in bad faith." This standard has been used repeatedly by courts over the years (e.g., see *Marquez v. Screen Actors Guild, Inc.*, 525 U.S. 33, 44 [1996]). As Professor Rubinstein has stated, "under Vaca, a union's actions are considered arbitrary if they are so far outside a wide range of reasonableness as to be considered irrational. A union's actions are deemed arbitrary if they fail to take a basic and required step under the collective bargaining agreement. Union action, or lack thereof, may also be 'arbitrary' if it results in an 'egregious disregard for the rights of union members.'"[24]

The employee does not have an absolute right to take a grievance to arbitration; this decision is left to the union. Today, some collective bargaining agreements may allow individual employees the right to file arbitration, at the expense of the employee and not the union, in the event that the union declines to take a grievance to arbitration. The intent is that the decision-making process be left to the union in choosing whether to take cases to arbitration; if the courts did not places limits on this, then the union, fearing legal challenges, would have to take every case to arbitration at a cost to the union and the employer.[25] In the case *Vaughn et al. v. Air Line Pilots Association International et al.*, No. 08-4173-CV (2nd Cir. May 14, 2010), the court referred to the duty of fair representation of all employees as "highly deferential, recognizing the wide latitude that [unions] need for the effective performance of their bargaining responsibilities." The NLRB has determined that unfair representation charges generally fall into the following four areas:

- *Improper motives*. The union cannot decide to process a grievance based on the person's age, race, gender, or the member's current view of the union.
- *Arbitrary conduct*. The union must make a good-faith effort to determine the merits of the grievance before deciding to process the grievance of dismiss the grievance.
- *Gross negligence*. Total disregard for the person rights, being a full union member or an agency fee–paying person.
- *Union conduct after filing*. The union has an obligation to process grievances with merit to a reasonable conclusion.[26]

The labor union has a responsibility to review current union practices in regard to the representation of its members. This should be an ongoing process, incorporating legal advice from a labor attorney, to protect the union from charges of failing to provide fair representation to its members. This may start with an annual review of the unions' bylaws and constitution to ensure compliance with handling of members' grievances. The union should provide written notice to the members outlining their rights to file their own grievance. It is generally in the best interest of the labor union to file a grievance and process it through all steps outlined in the collective bargaining agreement, unless it has been determined by a full investigation by the union that the grievance is meritless. The union should monitor and ensure that grievance deadlines are maintained by each party and whether needed extensions can be obtained and often granted by the other party. It is recommended that the union provide the grievant with a detailed written response to the process of the grievance and any developments. When applicable, the union should allow the grievant the opportunity to present evidence and testimony at the different levels of the grievance process.

As a former union member and union officer, I have at times found it challenging to put aside my personal opinions regarding a grievance. Even though the union collectively may not agree with the grievant, it still has the responsibility to ensure fair representation. This does not mean that a union should allow its members to file frivolous grievances to prevent being charged with lack of representation. Before the union decides to take a grievance to arbitration, the grievant should be allowed to make his/her case to the union—in front of the entire union membership (in a small local), to a grievance committee, or through some other means.

Regardless of the process, the union should ensure that it has a process in place to address whether grievances will be taken to arbitration. The union may want to consider the following questions in evaluating a case:

- Does the grievance have merit?
- How important is this issue to the union?
- What is the likelihood of winning/losing the case at arbitration?
- Could the case possibly be settled before the arbitrator issues an award?
- What extent does local politics play in the process?

In the event that the union has determined that it will not file for arbitration, written notice should be provided to the grievant outlining his/her right of appeal. In the event that the union has decided to pursue arbitration, the union should make it clear to the grievant that the union retains the right to withdraw the arbitration case.

Discussions between labor union officials and members are generally not considered *privileged* communication and therefore are not held as confidential; however, when conferring with an attorney, the communication is considered confidential. A significant portion of union business is conducted by non-attorneys behind closed doors; the union may represent the grievant at arbitration without the advice of counsel (in some cases, arbitration communication is privileged between lay union members). The courts recognize attorney-client privilege on communication when the union has consulted with an attorney on a case; however, they are reluctant to extend this privilege to lay union representatives. The State of Illinois has legislated labor relations privilege to labor leaders and members through 735 Ill. Comp. Stat. 5/8-803.5 (2007), which provides a shield to any union official who "in attending to his or her professional duties or while acting in his or her representative capacity" will not be compelled to disclose privileged communications; the state law clarifies that conflicting federal or state law will control. The Supreme Courts of Illinois and New Hampshire (926 A.2d 280, 283) have approved of state administrative decisions on privileged communications within the limits of employer and employee. In the public sector, the State of New York considers privileged any communications between labor leaders and members in respect to collective bargaining and administrative proceedings.[27]

Many courts and state labor administrative agencies have recognized that labor unions have a duty to fairly represent their members. The union has sole responsibility for preventing charges of failure to provide fair representation. However, the communication between the labor union and its members may not always be considered as privileged.

Past Practice

Sources of grievances are not limited to the written articles of the collective bargaining agreement. Labor or management may argue that a prior practice creates a binding condition of employment. A common misunderstanding is that a past practice has merit because of some conduct performed even once or just a couple of times. However, past practice involves much more. All too often, union members will sit back and throw past practices around as an implied agreement with many issues, in disharmony with the spirit of the legal meaning of past practice. Lacking a full understanding of the intent and merit of a past practice can have a chilling impact on the labor relations process.

The U.S. Supreme Court has established the standard for the inclusion of implied agreements. The implied agreement may be an agreement that the employer and the union have been allowed to engage in without disagreement for an extended period of time that has not been reduced to writing. The prior practice may clarify ambiguous contract language.[28] When the collective bargaining agreement is silent on an issue and there has been a standing practice, that constitutes a past practice.

Further, a past practice can be the result of a general or vague article in the contract. When the generalization or vagary leads to the conduct being carried out differently each time, this does not meet the burden for a past practice; however, when that generalization or vagary leads to conduct that is performed the same way over an extended period of time, this may rise to the benchmark of a past practice. Hence, if management or labor's conduct has taken place only one or a couple of times, then the burden for a past practice has not been met. The courts and arbitrators have not established a time frame or number of occurrences to meet the burden for past practice; therefore, the criteria will vary from arbitrator to arbitrator.

Supreme Court Justice Douglas stated, in the case *Steelworkers v. Warrior & Gulf Co., 363 U.S. 574*, "The labor arbitrator's source of law is not confined to the express provisions of the contract, as the industrial common law—the past practices of the industry and the shop—is equally a part of the collective bargaining agreement although not expressed in it." As attorney Matthew Franckiewicz said, "if you need to prove the existence of a historical practice, you need to find evidence that the conduct was consistent, clear, longstanding, and repeated, and betokens mural acceptance by both parties as the proper course."[29] As a general rule, to evaluate a claim of past practice, one should consider the following (fig. 6–5):

- *Longevity.* How long has this practice been in place?
- *Repetition.* How many times has this practice been conducted?
- *Consistency.* Has the practice been conducted the same way?
- *Knowledge.* Did labor and management have prior knowledge of the practice?
- *Acceptance.* At any point, did labor or management protest the practice?

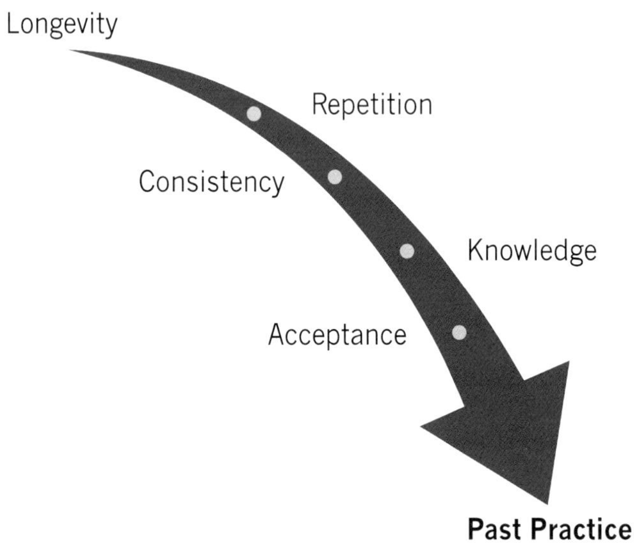

Fig. 6–5. Areas to consider when evaluating a possible past practice

After evaluating the merit of a past-practice grievance, the next logical process is to determine the type of practice. There are three categories of practices:

- *Conflicting practice.* A practice that is currently being performed is in violation of the language of the collective bargaining agreement. As these are by far the hardest cases to prove, most arbitrators will tend to rule in favor of the written contract language.
- *Clarifying practice.* A section of the collective bargaining is general or ambiguous, and labor and management cannot agree on the same meaning. This is the strongest form of past practice. The practice will need to be backed up with bargaining-table conversations and notes from the past.
- *Independent practice.* The issue is not covered in the collective bargaining agreement. An example would be benefits that employees have received on a regular basis for an extended time under mutual agreement but have

never been reduced to writing. In this case, management cannot usually be changed or altered without showing one or all of the following:

- A significant *change* has taken place to alter the original conditions of the practice
- A pattern exists of employee abuse of the practice
- Due notice was served on the union at the bargaining table that the practice will *change/terminate* under the new collective bargaining agreement

As a word of caution, it is best to determine the extent of your state's bargaining laws when it comes to changing past-practice issues. As a general rule, change and abuse should be discussed with the union first at the bargaining table, before implementing the change/termination of the practice, to ensure that you are not breaking any bargaining laws in your state.

The arbitrator will often use past practice as a beacon in interpreting the labor agreement between labor and management. On the basis of interaction or lack of action, the arbitrator may be able to clarify an ambiguous or general section in the contract. The arbitrator may be able to provide an enforceable condition to an agreement outside the collective bargaining agreement. Hence, past-practice issues are grounded in the contract and exist outside the contract.

The term *past practice* is often thrown around the fire house with ease, yet few can detail the exact limitations and extent of past practice. The firefighter and supervisor should have a clear understanding of the department's past practices. This can represent a learning curve for the supervisor or fire chief who is hired from outside the department and is unfamiliar with the department's past practice. In such a case, it is recommended to seek input from members within the union and from management to ensure you are on solid footing before taking action. As the outsider, you may have a strong opinion on the past practice; however, it is best to be respectful of the department's history and tradition. The past practices are unique to the individual fire department and the group of employees. What serves as a past practice in one department is not a past practice in another.

Consolidated Rail Corporation v. Railway Labor Executives Assn., 491 U.S. 299 (1989)

For a case review, see http://supreme.justia.com/us/491/299.

Final Thoughts

Just as in any relationship between humans, labor and management may have disagreements. The disagreement may result in a bargaining impasse. During the contract administration stage, labor and management each have a responsibility to uphold the collective bargaining agreement. Labor and management may occasionally disagree on the interpretation of contract language. A suitable means should be available to each party to settle any disagreements over the collective bargaining agreement. The goal of the dispute resolution process should be to allow an orderly, impartial, and quick resolution to a dispute.

This chapter has explored the grievance process, labor arbitration, mediation, and alternative dispute resolution procedures. The labor arbitration process was explained in great detail, because of the weight that it holds in the public sectors—especially in the fire service, given that fire unions cannot strike. Labor and management need to monitor the root causes of grievances to see whether patterns emerge that should be addressed. These include additional resources that are needed, contract language that needs to be clarified, labor and management hierarchical dysfunction, and/or poor communication. The human resource division can consequently suggest areas of improvement by way of training and education.

As a result of employees' paying dues in a labor union, the labor union has a duty to fairly represent its members. Union organizations are not allowed to pick who they will and will not help. The employee and the union each have a degree of responsibility. The following should be considered when evaluating a duty of fair representation: improper motives, arbitrary conduct, gross negligence, and the union's action after filing.

One labor relations topic that is often discussed around the fire hall is past practice. A significant number of grievances are based on or revolve to some extent around past practice. Regrettably, this term is often abused and is misunderstood by many. In considering a past-practice claim, one should take the following into consideration: longevity, repetition, consistency, knowledge, and acceptance.

Key Terms

- Alternative Dispute Resolution (ADR)
- Alternative Dispute Resolution Act
- Ambiguous contract language
- Arbitrary conduct
- Arbitration
- Award
- Clarifying practice
- Conflicting practice
- Duty to sign
- Ex parte hearing
- Fair Representation
- Federal Arbitration Act
- Federal Service Impasse Panel
- Grievance
- Grievance procedure
- Grievant
- Gross negligence
- Independent practice
- Interest disputes
- Improper motives
- Judicial review
- Longevity
- Online dispute resolution
- Package
- Past practice
- Permanent arbitrator
- Postal Reorganization Act
- Privileged communication
- Renew notice
- Rights disputes
- Sick-out
- Steelworkers' Trilogy
- Taft-Hartley Act
- Textile Workers v. Lincoln Mills Case
- Tripartite panel

Exploring Issues on the Web

American Arbitration Association: http://www.adr.org/ (Providing services to individuals and organizations who wish to resolve conflicts out of court)

CPR Institute for Dispute Resolution: http://www.cpradr.org/ (Nonprofit organization based in New York City, offering Podcasts on various topics. CPR's mission is to spearhead innovation, to promote excellence in public and private dispute resolution, and to serve as a primary multinational resource for avoidance, management, and resolution of business-related and other disputes.)

Federal Mediation and Conciliation Service: http://www.fmcs.gov/internet/ (Independent agency, founded in 1947, whose mission is to preserve and promote labor-management peace and cooperation. Headquartered in Washington, D.C., with two regional offices and more than 70 field offices, the agency provides mediation and conflict resolution services to industry, government agencies, and communities.)

Interagency Alternative Dispute Resolution Working Group: http://www.adr.gov/ (Lists resources to assist government employees and private citizens who use alternative dispute resolution in federal disputes and to support federal agencies in developing alternative dispute resolution programs.)

National Arbitration Center http://www.lawmemo.com/arb/award/default.htm (Free site used to search full-text awards submitted by arbitrators) National Mediation Board http://www.nmb.gov/ (Independent agency playing a pivotal role in facilitating harmonious labor-management relations in the railroad and airline industries.)

References

1. Carrell, M., and C. Heavrin. 2010. *Labor Relations and Collective Bargaining*. 9th ed. Upper Saddle River, N.J.: Pearson Prentice Hall.
2. Fossum, J. 2009. *Labor Relations*. 3rd ed. New York: McGraw-Hill Irwin. P. 463
3. *Stanton v. Delta Airlines*, 669 F.2d 833, 838 (1st Cir. 1982).
4. Bohlander, G., and S. Snell. 2010. Managing Human Resources. 15th ed. Mason, Ohio: South-Western Cengage Learning. P. 641.
5. Zander, A. 2007. A union view of collective bargaining in the public sector. In *Public Personnel Administration and Labor Relations*, ed. N. Riccucci. Armonk, N.Y.: M.E. Sharpe
6. Jackson, S., R. Schuler, and S. Werner. 2009. *Managing Human Resources*. 10th ed. Mason, Ohio: South-Western Cengage Learning. P. 536.
7. Elkouri, F., and E. Elkouri. 1999. *How Arbitration Works*. 5th ed. Washington, D.C.: Bureau of National Affairs.
8. Nolan, D. 1998. *Labor and Employment Arbitration*. St. Paul, Minn.: West Group.
9. Elkouri and Elkouri 1999, p. 101.
10. Bennett, S. 2002. *Arbitration: Essential Concepts*. New York: American Lawyer Media.
11. Elkouri and Elkouri 1999, p. 128.
12. Nolan 1998.
13. Ibid.
14. American Arbitration Association. 2007. Drafting dispute resolution clauses: A practical guide. http://www.adr.org/si.asp?id=4125 (accessed September 10, 2010).
15. Elkouri and Elkouri 1999, p. 20.
16. Ibid., p. 116.
17. Wilkinson, J. 2009. Arbitration contract clauses: A potential key to a cost-effective process. *Dispute Resolution Magazine*. 16 (1): 9–12.
18. Portnoy, R. 2010. Embracing the alternative: *Cable Connection, Inc. v. DirecTV, Inc.* puts the alternative back into alternative dispute resolution. *New England Law Review*. 44: 991.
19. *United Steelworkers of America v. Warrior & Gulf Navigation Co.*, 363 U.S. 574, 581-582, 34 LA 561, 564 (1960).
20. Budd, J. 2008. *Labor Relations: Striking a Balance*. 2nd ed. Boston: McGraw-Hill Irwin. P. 358.
21. Witney, F., and B. Taylor. 1996. *Labor Relations Law*. 7th ed. Upper Saddle River, N.J.: Prentice Hall.

22. Jackson et al. 2009.

23. Budd 2008, p. 351.

24. Rubinstein, M. 2009. Duty of fair representation jurisprudential reform: The need to adjudicate disputes in internal union review tribunals and the forgotten remedy of re-arbitration. *University of Michigan Journal of Law Reform*. 42 (3): 517–55.

25. Outten, W., R. Rabin, and L. Lipman. 1994. *The Rights of Employees and Union Embers*. Carbondale, Ill.: Southern Illinois University Press.

26. Jackson et al. 2009, p. 537.

27. Rubinstein, M. 2008. Is a full labor relations evidentiary privilege developing?. *Berkeley Journal of Employment & Labor Law*. 29 (1): 221–66.

28. Cooper, L., D. Nolan, R. Bales. 2001. *ADR in the Workplace*. St. Paul, Minn.: West Group. P. 324.

29. Franckiewicz, M. 2009. How to win your arbitration case before the hearing even starts. *Labor Law Journal*. 60 (3): 115–20.

DISCIPLINARY ACTION 7

Learning Objectives

After reading this chapter, you should be able to:

- Define the steps in conducting an employee complaint.
- Recognize the importance of conducting a threat assessment.
- Explain employee rights as they pertain to discipline and termination.
- List the important elements of a separation agreement.
- Explain how due process applies to public sector discipline and termination.

Disciplinary Action

Many public sector employees have been classified as at-will employees, meaning that they can be disciplined and discharged for any reason or for no reason. However, as a result of collective bargaining agreements, many of these same employees are now covered under laws pertaining to discipline and discharge and have some level of protection under the collective bargaining agreements, civil service laws, state laws, and/or municipal policies and procedures.[1] Discipline is a personnel issue, with an impact on both union and nonunion personnel. From a personnel management standpoint, discipline and termination actions taken by the employer are subject to review by the grievance process, civil service, and courts and therefore can be reversed.

This chapter will take you from the starting point, when a department receives a complaint about an employee issue, through the steps taken to ensure that the disciplinary action is on solid legal and ethical footing. Emphasis will be placed on the protection of the rights of the employee who is being disciplined. Public sectors employees are offered protection under the 14th Amendment when it comes to due process. Given the significance of discipline and termination actions toward labor relations and personnel management, it is essential to have a clear understanding of how the process works and the steps that can be taken in advance to maintain a solid labor management relationship and minimize the legal risk associated with disciple and termination.

The Starting Process

Every effort should be taken to minimize the risk that a disciplinary action or termination may be overturned. Any good labor union will put the fire department's investigation on trial in an effort to divert attention from the actions of the disciplined firefighter. It highly recommended that discipline investigations begin immediately; any delay in starting an official investigation may later be used to show that the fire department did not see the matter as serious, and the investigation may be inherently flawed by the delay. Fire chiefs may be faced with a long list of why they cannot start an investigation (too much work, large incident they are overseeing, vacations, between shifts, etc.); however, there is no excuse for delaying any investigation into a possible disciplinary action, as the sooner the investigation is started, the greater will be the chance of obtaining accurate answers and admissions. Equally important is finishing an investigation. There must also be consistency in the application of the disciplinary action; inconsistency can result in cases being overturned.[2]

The chief must consider, based on the complaint, whether an employee(s) needs to be placed on administrative leave, pending the investigation. This may prevent the employee from tampering with evidence, reduce the chance of interfering with the investigation/witnesses, or reduce the chance of retaliatory action on the accuser. Immediate administrative leave is generally on a paid basis until the end of the investigation or criminal proceeding. The investigating officer should interview any person who may have information pertaining to the incident/issue. The investigation should be conducted without bias or subjectivity. Investigation into an employee complaint should include the following:

- Promptness (the investigation starts immediately)
- Thoroughness (interviewing all witnesses, reviewing all facts, reports, and storing all evidence collected)
- Fairness/impartiality
- Accurate (all reports should be typewritten and checked for spelling/grammar, in addition to accurate facts about the case)
- Confidentiality
- Demonstration of a systematic investigation

Threat Assessment

As a first-line supervisor or fire service administrator, it is important to have a full understanding of the principles of discipline, because at some point in your career, you will be faced with taking action against one of your personnel (fig. 7–1). As a firefighter (employee), it is equally important to know your rights to ensure that a supervisor and/or municipality does not violate your rights. When it comes to personnel and discipline issues, many first-line supervisors and administrators hope the problem will go away on its own; similarly, most employees do not find enjoy being called into a disciplinary meeting, wondering whether they will be terminated. The heightened emotions on each side generally makes disciplinary meetings very tense. As a result of the high anxiety involved in discipline meetings and especially termination hearings, the first order of business is to assess the safety of the organization and the people in the meeting/hearing by conducting a threat assessment.

Fig. 7–1. Today's fire service administrators must have a clear understanding of disciplinary actions and legal requirements to ensure that the employee rights are not violated.

As a general rule, employees often know when they are about to be disciplined or terminated—only in very rare cases does this action come as a total surprise. They may have the means to seek revenge on the person taking the action. The past and current behavior of any employee about to be called into a disciplinary hearing should be assessed. Therefore, a security plan should be developed in case the meeting turns violent, and the hearing should be held in a remote secured room.

Fire service administrators should partner with the human resource department and law enforcement in conducting a threat assessment before holding a disciplinary or termination hearing. It is recommended that the local law enforcement agency be notified that a termination hearing will be conducted, so that law enforcement officers can be in the area and available for assistance; the need for a coordinated effort with law enforcement will depend on the significance of the incident(s) resulting in the discipline and/or termination. Will the employee need to be advised of a *no-trespass* order on municipal land without being escorted by local police? One recommendation is to hold a discipline/termination hearing away from the fire station or administration building, in the city or town hall. Regardless of the building in which the hearing/meeting is held, a threat assessment must be conducted for the people involved and the occupants of the building. A plan should be made in case the building needs to be locked down because of an incident in the meeting/hearing. How will the message from within the meeting be transmitted to people outside the room and how will the lockdown be announced for the building? The fire service administrator may be too close to the individual or incident to properly identify threats; thus, having a second set of eyes evaluating the threat is best. Further, by collaboration with human resources and law enforcement, you minimize the risk of abusing the system and merely embarrassing an employee.

Workplace aggression and violence

Workplace aggression and violence have become predominant in the news media of late, with no relief in sight. Often, workplace aggression and violence results from the employees' real or perceived inequities. When employees feel (correctly or not) that they are not being treated fairly and respected, their response may range from the theft of copying paper to the destruction of department property/equipment. Such behavior is frequently seen after disciplinary action, pending force reductions, being passed over for a promotion, and termination. The department administration should all be made aware of employees who may exhibit signs of aggression and/or violence (fig. 7–2), as left unchecked, this behavior can poison others within the organization.[3] This is a great example of the human resource triad (see chap. 8) working together to prevent workplace aggression and violence.

Fig. 7–2. Labor and management must work to prevent workplace aggression and violence in the fire service.

Education is one of the keys to dealing with workplace aggression and violence. Employees should be made aware that they are to report incidents of aggression or violence and assured that the department handles such matters seriously. The fire department must have a zero-tolerance policy for retaliatory actions against any person(s) reporting an incident of workplace aggression and violence. Policies should also include a section on filing false claims or taking retaliatory action against another employee.

In the event of an exigent circumstance, the department must take immediate action to ensure the safety of the work environment. Whether this means evaluating a room or a section or the entire station, it should be ensured that all employees are safe. Do not be lulled into complacency; nothing could be more dangerous. When a person is in a high stage of stress or despondent, he or she is capable of doing anything—yes, even hurting or killing brother and sister firefighters. If the reporting party cannot be provided with a safe work environment, the department may want to consider offering him/her a day off with pay to allow the department time to assess and take corrective action.

If the department removes an employee from the property because of aggression or violence, the law enforcement agency must be involved, and if needed, the employee should be advised of a no-trespass order on municipal land. Employees should not be allowed to gather their personal belongings from the locker room or anywhere else; instead, ask what is needed and send a police officer and a supervisor to gather the employee's belongings. It is recommended that the municipality prepare a written list of items turned over to the employee and have the person sign for them. The same method should be used if the employee is requesting additional personal items later.

Depending on the department's policy or the municipal handbook, there may be a need to conduct an *inventory search*, to identify all of the items at that point in time, in the employee areas, such as locker rooms and storage areas. The use of still photography or videotaping is recommended, and a list of the items found should be prepared in writing. If the items are in an unsecured area, it is recommended that they be locked up and secured, to ensure that they remain safe. Before taking any such action, seek the advice of an attorney. This is a policy that should be developed now, before an incident takes place. Your local law enforcement agency will be able to help you prepare such a policy.

Employee Rights

When discussing disciplinary action and termination issues, it is important to recognize that employees have rights, granted by the courts, legislation, or the employer. The basis of employee rights is that employees are guaranteed fair treatment when it comes to their employment status. It is important to point out, as attorney Will Aitchison does, that "volunteer firefighters are generally not 'employees' for purposes of coverage under Title VII of the Civil Rights Act, and thus cannot bring employment discrimination or harassment lawsuits under Title VII, nor are they typically covered by the Fair Labor Standards Act."[4] Generally, volunteer firefighters have little to no rights when it comes to state and federal laws, unless they are specifically addressed in the laws. When discussing employee rights, one must first define who is considered an employee.

A common problem arises when an employee claims that his/her rights to privacy have been violated by the employer. For example, firefighters may feel that they have an expected level of privacy when sending/receiving e-mail communication via department computers, of freedom from searches of lockers issued to firefighters for storage of personal belongings, and of privacy when using a personal cell phone on the job. Many municipal agencies now have issued warnings and notices that the e-mail system is open to public disclosure and/or to review of use/abuse by the municipal agency. Often the warning appears when a firefighter logs on to the computer system that there is no expectation of privacy.[5] It is strongly recommended that the fire service administrator reviews the case with the legal department before taking any disciplinary/termination action to ensure that the employee's rights have not been violated or are not about to be violated.

Much of the supervisor's time will be spent addressing personnel issues, such as unacceptable behavior resulting in some level of disciplinary action. The level of discipline used by the first-line supervisor may be limited to simple oral and written reprimands for firefighters under his/her command. Any disciplinary action that involves suspension (with pay or without pay), demotion, or reassignment may have to come from the fire chief or the municipal leader. It is important to be aware of the limits of your rank when it comes to a disciplinary action.

When it comes to personnel issues, the good news is that generally an oral or written reprimand will be sufficient to correct most problems. The key ingredient to a successful correction is good communication between the supervisor and the employee, respect between the two parties, and a path that supports success for the employee.[6] When supervisors attempt to embarrass an employee or set a path for the employee to fail, it is then that employees will often seek relief through the grievance process, civil service, and the courts. The supervisor's goal should be to always conduct a thorough analysis of the cause of the unacceptable behavior and provide a positive action plan to remedy the issue.

When considering disciplinary actions, one must distinguish between the types of unacceptable behaviors. In the case of *culpable behavior*, the employee is at fault, as when calling in sick on the opening day of fishing season or defacing employer property. *Nonculpable behavior* is when the firefighter is not at fault, as when calling in sick because the car will not start or because of inability to understand and perform simple directions or assignments from a supervisor. The level of discipline imposed may reflect whether it is in response to culpable or nonculpable behavior. The culpable behavior may elicit a stronger disciplinary action, whereas the nonculpable behavior may be corrected with awareness and training. When it is difficult to determine whether an employee's behavior is culpable or nonculpable, it is recommended that the behavior be classified as nonculpable.

The *reprimand* is a stronger discipline than a *warning*. Note that, in the case of culpable behavior, reprimands are issued; in the case of nonculpable behavior, warnings are issued. When time off from the job is deemed appropriate, it is called a *suspension* when in response to culpable behavior (e.g., "Due to your misconduct") and a *forced leave of absence* when in response to nonculpable behavior (e.g., "Due to your inability or failure"). When separation is deemed appropriate, it is called a *discharge* when in response to culpable behavior (e.g., "Due to your misconduct") and a *termination* when in response to nonculpable behavior (e.g., "Due to your inability or failure") (table 7–1).

Table 7–1. Culpable and nonculpable behaviors

Culpable	Non-Culpable
Verbal Reprimand	Verbal Warning
Written Reprimand	Written Warning
Suspension	Forced Leave of Absence
Discharge (for just cause)	Termination (inability or failure to meet the duties

Two-way communication is critical in the discipline process. This means it is not the supervisor doing all the talking; as discussed in prior chapters, the supervisor needs to allow the employee tell his/her side of the story. More often than not, an employee who is allowed to communicate openly with a supervisor during the disciplinary process perceives the process as being fair; conversely, when an employee is not offered the chance to speak or has the feeling that he or she is not being heard, the employee will most likely perceive the process as unfair and resent the disciplinary process. Generally speaking, it is best to sit down with the employee and explain the facts that have led to this meeting. Then, allow the employee ample time to refute the merits of the claim against them. After the employee has a chance to refute the merits and ask questions, it is up to the supervisor to explain what action plan will be taken. This may mean that additional information may be needed and no action will be taken today. Based on the responses from the employee, the supervisor may have gained a clear idea as to what action is needed. When the supervisor has obtained adequate information surrounding the incidents/issues, he/she must clearly explain what the action plan will be. The action plan will explain what steps will be needed to turn the unacceptable behavior into acceptable behavior. The consequences if the unacceptable behavior continues must be clearly stated. Be sure to solicit follow-up questions: "Do you have any questions about the action plan?" and "Do you understand what will take place in the events the unacceptable behaviors does not change?" It is best to end the meeting with a recapitulation of the meeting to ensure that everyone understands what has been discussed.

In the case of a disciplinary action, it may be best for the supervisor to prepare a written account of the meeting and to ensure that all parties clearly understand the corrective action needed and, if not corrected, what the consequences will be. The employee would then be given a copy of the document and asked to sign acknowledging that they received a copy. If they do not sign it, you could just write on the department copy that the employee refused to sign it. In one case, I had an employee refuse to sign a discipline sheet, so I asked if he would just write on the sheet that he is refusing to sign it. He did, which could be used to show arbitrators that the employee received the sheet. Hence, the employee had prior knowledge what actions would be taken in the event that the behavior did not change. This goes in line with the due-process principles that will be discussed later in this chapter.

Simple two-way communication can go a long way toward defusing many emotional issues that the supervisors will encounter. The two-way communication process should be used by supervisors every day and not just when they are faced with a problem.

Personnel Policy

In theory, when a fire department has an established personnel policy that is updated on a regular basis, disseminated to all employees, and uniformly applied, the need for discipline will be minimized. When employees and supervisors understand why the policies are in place, they are more willing to follow them and support the policies. When most fire departments run into trouble with discipline, it is the result of one or a combination of the following:

- Policies not being followed the same way for all employees
- Employees did not have a clear understanding of what constitutes acceptable versus unacceptable behavior
- Poorly documented incidents (no report or a minimal written report)

One danger that all fire departments face is the lack of not conducting a *policy update/edit* (fig. 7–3). This entails more than just updating current policies; rather, it is intended to reevaluate the need for current policies. This review should include the employee handbook. Some fire departments find out the hard way that the policies that they once thought would protect them in a lawsuit are now being used against them. It is recommended that the municipal human resource department track how many infractions have been issued as a result of each policy; this will show how the policy is being used or not used and the potential for disparage application. In regard to the tracking of infractions to policies, one attorney has stated, "But keep it only for the purpose of getting legal advice so it's protected by the attorney-client privilege and it won't be discoverable in an employee lawsuit."[7] If a department finds a policy that has been in the handbook for years over which time it has not been followed, now is the time to remove it. Having policies that are not followed will work against the department's interests and may open the department up to further litigation. The fire department handbook should not include any policy by which the department is not going to abide. As Diane Cadrain has stated, "rather than shielding the employer from liability, the old policy actually may invite litigation if a manager decides to enforce it after it's lain dormant for years. That's because real world practice often ends up trumping languishing handbooks policies in court."[8] Outdated and sporadically enforced policies are a gateway for employees to file a claim of discrimination under state and federal laws. If the employee can show that the action is based on unequal treatment on the basis of sex, race, age, disability or other protected classification, then they may file a claim of discrimination.

Fig. 7–3. Fire departments must ensure that department personnel policies are updated, disseminated, and reviewed with all employees on a regular basis, to ensure that employees are aware of acceptable and unacceptable behavior.

When firefighters see that others are treated differently or that it takes a long time for discipline issues to be addressed, the organization, morale, and the case are affected. The labor arbitrator and the courts will often review the fairness of the disciplinary and/or termination action taken by the employer. Therefore, your fire department discipline/termination procedures must be fair and equitable for all employees. Further, fairness has a positive effect in many areas of the fire department—namely, heightened employee commitment and enhanced satisfaction with the job, supervisors, and administration.[9]

How the disciplinary action is handled may affect the labor-management relationship. The union will want to ensure that the employees' rights have not been violated and that the disciplinary action was in accordance with the standards established by the department. In fact, the number-one issue disputed in front of labor arbitrators is termination, with the second leading issue in front of arbitrators being discipline.[10]

One of the biggest challenges fire officers face is the possibility of disciplining friends they have known and worked with for years. On a day-to-day basis, it may be great to supervise the buddies you have worked alongside of over the years; however, when things turn bad, they are often compounded by this prior relationship. If the supervisor sticks his/her head in the sand hoping the problem will go away, then the supervisor will lose respect from the troops. If the supervisor takes a hard-nosed approach to set the tone that he/she will not tolerate this behavior, the troops may resent the supervisor for being poisoned by the promotion. Finding that balance of being fair but firm is not an easy task and ultimately comes from experience in grade.

Another concern for supervisors is how to respond to a procedural injustice, when a disciplinary action that they have taken gets overturned by a higher authority. In such a case, a supervisor may exhibit abusive behavior against the employee or the people who testified against his/her actions. It is recommended that the supervisor who loses a disciplinary case be advised of any retaliatory actions that he/she may take against people involved in the case. Simply making a person aware of this potential retaliatory action may be enough to prevent such actions.

There is no easy school or class you can take to make the best call every time. First-line supervisor training should include sections on how to discipline employees. Senior fire officer programs should spend a great deal more time on discipline issues including termination. They need to know how to investigate a discipline issue, prepare a written case, and present their recommendations to a hearing officer. Chief officer training on disciplinary issues should be required for both the fire chief and the municipality human resource officer. This process will ensure that the human resource officer is working in coordination and under the municipal government laws and regulations. Thinking outside the box, the labor organization and the chief officers could attend a combined training seminar on discipline issues to ensure that each is receiving the same information.

Discipline has long been viewed in terms of "us against them," and today many fire departments are moving toward a partnership with discipline issues. When administration involves the labor union and groups of employees in establishing and reviewing department policies on discipline, the process is better received. When employees understand the need for policies and are involved in the process, they are more inclined to support the department policies. Many issues and concerns that individual employees as well as the union may have surrounding disciplinary issues can be resolved well in advance. This can save time, resources, and money in the long run for everyone involved. Most important, the process may preserve the labor relationship between management and the union. Labor and management should work together to resolve any disciplinary procedural issues well in advance rather than wait until an employee is facing a discipline action.

A Warning about Warnings

Many people know that the issuance of warning for disciplinary action is a standard action. This is a growing concern in the litigation and human resource fields that now raises a concern with the issuance of written warnings. It goes without saying that a progressive discipline program lays out the steps that the employer and the employee have taken in correcting the issue. In fact, the progressive discipline approach provides a solid paper trail of evidence, if conducted correctly. This approach can also minimize the fire department's exposure to discrimination claims. There is growing evidence that the warning provided in a progressive discipline program can leave the department at risk for unlawful retaliation charges by the employee. The U.S. Equal Employment Opportunity Commission (EEOC) in fiscal 2009 reported that the two most frequently filed charges with the EEOC are race (36%) and retaliation (36%); furthermore, as the rate of claims for retaliation with the EEOC has been steadily increasing over the past several years, it would behoove fire departments to reexamine their current disciplinary policy.[11] At issue is that employees can now claim that discipline (e.g., a warning) comprises a retaliatory action.

As the boxed hypothetical case indicates, issuing a written warning can leave the department open to retaliatory claims by employees. If Captain Hill continued to issue written warnings to Firefighter Brown, it might decrease the risk of discrimination while opening the door for claims of retaliation claims. Fire department administrators must conduct a risk-benefit analysis to compare the benefits of issuing employees written warnings against the associated risks. This is not about eliminating a risk; rather, it is about determining the least amount of risk when it comes to progressive discipline. More paperwork and more documentation is not always the best avenue when dealing with discipline issues.

The fire department's best defense to a retaliation claim is creating and following a set of policies or procedures that are nondiscriminatory. The department policies and procedures need to be disseminated to all employees on an annual basis. Providing copies of policies and having the firefighters sign that they have received the policies and procedures is not ample. The department should be also conducting training session on the policies, to show the employees and supervisors what are the acceptable behaviors for the department. It is recommended that the training sessions identify both what unacceptable behavior is and what will be the consequences for unacceptable behavior. By identification of the acceptable versus unacceptable behavior and of the consequences for unacceptable behavior, employees are less likely to exhibit an unacceptable behavior. Supervisors and fire service administrators should receive specialized training on the department policies and procedures and how to enforce them. By providing training and support to supervisors and administrators, the disciplinary actions taken will ensure consistent employee conduct as outlined in the policies/procedures and will lend itself to fair and nondiscriminatory treatment. In simple terms, the fire department needs to publish and follow a uniform method when dealing with employee issues. Having written policies is not a substantial defense; the organization must set and follow the behavior it has endorsed by way of the policies and procedures.[12]

Hypothetical Case

Firefighter Brown has been on the job for the past two years. Her performance had been fine for the first 20 months; only within the past 4 months has her performance declined. Firefighter Brown's supervisor is Captain Hill. During the past month, out of frustration with Firefighter Brown, Captain Hill tells her that he wants to fire her. Captain Hill comes into the fire chief's office to consult with the chief about possibly terminating Firefighter Brown. The chief reviews the personnel file for Firefighter Brown, only to find out that she has not received any progressive discipline while on the job and, in fact, on her one-year anniversary, she received an employee evaluation that indicated that she is performing all of her tasks. Based on the information in the personnel file, the chief informs Captain Hill to provide Firefighter Brown with written notice of her substandard performance and what is the standard she must meet. Captain Hill is not happy about this approach. However, he reluctantly prepares a written warning and presents the warning to Firefighter Hill.

One week later, the chief of department receives a letter from an attorney who represents Firefighter Brown. The attorney claims that Captain Hill's action is a product of sex discrimination, because other male firefighters are allowed to do the same without receiving any disciplinary action.

The chief immediately conducts an extensive investigation into the incident. The investigation finds that none of the other male firefighters had the deficiencies for which Firefighter Brown had been cited in her written warning. The chief speaks to Captain Hill about his investigation and the results.

Approximately one month latter, Firefighter Brown's attorney sends another letter to the chief of department, now claiming that since Firefighter Brown has filed her original complaint for discrimination and harassment, she has suffered retaliatory actions from Captain Hill and the fire chief. The letter goes on to state that Firefighter Brown has been asked to go out and shovel snow away from fire hydrants and has been declined training opportunities. All of which have taken place since Firefighter Brown filed her original complaint. Captain Hill and the fire chief must now demonstrate that the actions they took in this case are not retaliation.

You can see how a simple disciplinary action can lead to a major legal battle in a case like this. What can be learned from this example? What could have been done differently?

Termination Procedures

Terminating the employment relationship may be one of the riskiest actions taken by a fire chief, owing to the myriad of legal recourse available to terminated employees. Today, the courts are further defining the responsibilities of the employer and the employee. For example, previously, the call firefighter could be terminated for any cause or for any reason (except for illegal actions like disciplining or terminating a person because of race, sex, religion, age, or national origin); today, however, there may be an implied employment agreement based on the fire department's handbook or policies. In years past, the courts have been less inclined to review such a case; however, courts are now more inclined to review such a case based on individual employment contracts, collective bargaining agreements, civil service law, or public policy. There is no easy answer as to the level of protection that a public employee is afforded when it comes to discipline and termination; rather, a number of factors come into play.

Employee terminations are never easy on the person performing the termination, the employee being terminated, or the organization. Fire service administrators should take into consideration the legal and ethical challenges to terminating the employment relationship. Most municipal and fire service leaders focus on the legal aspect of the termination and give minimal consideration to the ethical obligation when terminating employees. How the termination is conducted will have a ripple effect throughout the fire department, moving out from its core values to the entire working environment. It should go without saying that fire service administrators should not be terminating a firefighter on the Wednesday afternoon before Thanksgiving. Termination should occur early in the week to allow the terminated employee time to seek legal advice and to ask follow-up questions. As Van Bogaert and Gross-Schaefer have pointed out,

> Even though an organization is on firm legal ground to initiate layoffs based on legitimate and compelling business necessary, maintenance of core values, decency, and respect for effected employees may be equally important. There is a wrong and a right way to handle involuntary terminations of employment, in addition to a long list of legal consideration. Proper termination management examines both ethical and legal challenges.[13]

When terminating an employee, the department should have prepared a well-documented case leading to the termination. Granted, some behaviors are so serious or heinous as to warrant immediate termination, and the employer may in such a case have limited documentation on the employee's behavior beyond the single incident. A serious or heinous infraction will still require a detailed investigation into the incident and must include the employee's side of the story. Until all of the facts have been compiled and reviewed by administration, the decision should not be made as to whether the employee can return to work. Serious incidents are, by definition, incidents that cannot be allowed to occur more than once.

It is recommended that employees receive ample notice, explanation of the unacceptable behavior, and a course of action to correct the unacceptable behavior. The written documentation should outline the following: each of the incidents that have resulted in the unacceptable behavior; request for corrective action by the employer; employee evaluations; and any employee responses to past corrective actions or employee evaluation. The theory behind the progressive discipline process is to identify unacceptable behavior and make the employee aware of it, as well as of the needed corrective action and how to achieve the corrective action. Many organizations will establish a time frame for obtaining the corrective action and the consequences if correction is not achieved.

In the case of at-will employment status, not all departments will immediately terminate the employee; instead, many will offer a process that allows the employee to correct the unacceptable behavior. As discussed previously, at-will employees who believe that they have been terminated for a discriminatory reason can file a claim with the state and federal government. All state and federal discrimination and equal opportunity standards supersede local policies. Many organizations with at-will employees have eliminated or minimized their policies on terminating at-will employees, because these documents have been linked to contract termination-without-cause clauses. For example, in the case of *Harper v. Healthsource New Hampshire, Inc.* (140 N.H. 770, 674 A.2nd 962) (1996), public policy, rather than discrimination, was at stake. Because the employer did not indicate why they terminated the doctor, the doctor was prevented from responding to the termination.[14]

Many would argue that the relationship between labor and management is more equitable today than in years past. Nevertheless, the employment relationship has become more complex.

There are three basic types of wrongful termination actions:

- Violation of public policy
- Breach of express or implied contract
- Implied covenant of good faith and fair dealing

When violation of public policy is alleged, the employee argues that the termination violated a public policy. In this case, each state court may have a degree or burden that must be met in order for the case to be reviewed by the court. Clearly established written public policy (i.e., constitutional provision or state statue) will more often than not succeed in court. For example, many states have written statutes that provide employees the right to serve on jury duty and that assert the employee cannot be terminated for not being at work while serving on a jury (providing that the employee served notice to the employer of the jury duty).[15]

In the case of an expressed contract, the terms and conditions are stated orally or in writing. In the case of an implied contract, the terms and conditions of the contract are inferred by the behavior and actions between the two parties. The courts will infer from the fire department's policies/guidelines/handbook and the behaviors/actions between the two parties or that are implied from law. Thus, if the fire department handbook details how employees will be terminated for certain enumerated infractions on the basis of policies/procedures, the employee could argue that the handbook was incorporated as part of his/her employment contract with the department.

Finally, some state and federal courts have said that an at-will employee cannot be terminated if the termination is unfair or in bad faith. In spite of the significance of this determination by the courts, they stop short of saying that at-will employee involuntary termination must be for just cause.[16]

Separation Agreements

Fire service administrators, who are not covered by a collective bargaining agreement, often will enter into a separation agreement at the termination of employment. The separation, or severance, agreement may offer the terminated employee certain postemployment benefits. Among these benefits may be a continuation of salary for a defined period of time. A significant issue today is the extension of health insurance for an extended period. Some fire service administrators will enter into an individual contract with the municipality, and the contract may detail the separation process if the municipality seeks to end the employment relationship.

Sample Separation Contract Article

In the event that the chief is not reappointed or the chief resigns following a formal suggestion by the city that he/she resign before the expiration of the then applicable term of employment, the city agrees to pay the chief a lump sum severance payment equal to the balance of any term of appointment, but in no case less than twelve (12) months salary and benefits as prescribed in the employment contract.

The intent of the separation agreement is to ensure that the administrator will receive a set payment and benefits when the municipality wants to end the employment relationship. This serves as a small incentive for the municipality not to ask the administrator to leave and as a cheap insurance policy that will provide pay and benefits while the administrator seeks other employment. The length of the payment period will range from one to several months, depending on what the employee and the employer agree on. An extended payment plan (e.g., of six to eight months) is based on the time frame for a chief to find suitable employment elsewhere and, given the political impact of the position, strives to prevent the municipality from simply terminating the relationship because they don't like the administrator. It is important to keep in mind that not all employee contracts contain this benefit. From a risk management standpoint, the municipality must look at the money that will be spent on the wages/benefits as compared to on a prolonged legal battle (cost of attorney, length of case, and possibility of damages). This is why, in many cases, the 6-months separation agreement package amounts to much less money. The separation agreements will often contain language that the employee release the municipality from any claims for employment, meaning that the employee will not file suit against the municipality over the termination. Some states have severance agreement statutes that must be referenced in the agreement, to serve as a valid release of a legal claim. As stated by Bruce Fried and Myron Fottler,

> To obtain a valid release for an age-discrimination claim, an employer must follow the ADEA (29 U.S.C. §§ 621-34) requirements: The employer must provide:
>
> 1. A written agreement,
> 2. Consideration,
> 3. Advice in writing to seek counsel prior to signing,
> 4. 21 days to consider before signing,
> 5. Seven (7) days to revoke after signing (29 U.S.C § 626(f)[1]).[17]

Because of the complexity of legal issues surrounding separation agreements, the municipality should seek the advice of a labor attorney. The employee should also seek legal advice from a labor attorney, to ensure that his/her rights are protected and that the agreement is enforceable in court.

Due Process

The collective bargaining agreement, individual employment contract, and some employee handbooks will establish the procedure for filing a grievance or some other means to allow the employee due process. Due process must be available to all employees. In addition, public sector employees are provided additional protection pertaining to due process under the 14th Amendment, Section 1: "No State shall make or enforce any law which shall abridge the privileges or immunities of citizens of the United States; nor shall any State deprive any person of life, liberty, or property, without due process of law."[18]

In terms of employment, the courts will look to see if the employee had an expectation of continued employment. If the employee can be terminated for *cause* or *just cause*, then the courts will generally see that the employee had a property interest in their employment. However, at-will employees can be terminated for any reason or for no reason and therefore would not have a property interest in their employment. It is common for new firefighters to be placed on a 6–12-month probation period, which indicates that the employer can terminate the employment at any point and for any reason (depending on how the probationary clause is written).

In some cases, during the final stage of a progressive discipline process, the employee and the employer may enter into a *last-chance agreement* that will limit the employee's right to appeal and/or grieve the actions (if there is a labor union, the union must also accept the last-chance agreement). Last-chance agreements are often used with alcohol and drug violations, allowing the employee to continue employment, providing that new conditions of employment are met; if the new conditions are not met as mutually agreed on, then immediate termination would result. In these cases, the employee would normally be facing termination, but the employer is allowing the employee one last effort to correct his/her behavior.[19] The union may support the use of a last-chance agreement, because it will allow the employee to retain his/her job and to change his/her behavior. The alternative is to not sign the last-chance agreement and allow the employer to move forward with terminating the employee, which could result in a prolonged and costly appeal process.

Sample Last-Chance Agreement

The following constitutes a last-chance agreement between ___(Employee Name)___, Local ____, and the ___(Town/City)___ in conjunction with the Town of __________, human resources division, and the fire chief.

The Department agrees to:

The employee agrees:

It is agreed by all of the parties that if the employee violates the last-chance agreement, or if there is continued violation of (___), the appropriate discipline shall be termination from his/her position. The department need prove only that the employee violated the above agreement(s)/rule(s). The arbitrator shall have no authority to modify the discipline. All parties acknowledge the waiver of the contractual due-process rights to the extent stated above. This last-chance agreement is in force and effect for two (2) years from the date of the grievant's signature on this agreement.

Authorizing signatures:

Date ________________ ___________(Employee)___________

Date ________________ ___________(For the Union)___________

Date ________________ ___________(For the Town/City)___________

Date ________________ ___________(For the Chief of Department)___________

The last-chance agreement is a written agreement between the employer and the employee, including the following: the terms and conditions the two parties mutually agreed to; the time frame; and limitations the last-chance agreement will have as far as appeal/grievance process. It is believed that employees subject to a last-chance agreement would not have any property interest in their employment. However, in the case of *Farahani v. San Diego Community College* 175 Cal. App. 4th 1486 (July 28, 2009), the California Court of Appeals held that a last-chance agreement between the college and a faculty member attempted to waive the statutory due-process rights of the Education Code, Title 3, Division 7, Part 51, Chapter 3, and was therefore deemed unenforceable; thus, it is recommended that employer and employee each seek legal advice on any last-chance agreements that attempts to circumvent due notice and due-process hearings/appeals.

Public sector employees at the local level derive their employment interest from state law. The state labor laws will define whether the employee has a property interest in continued employment. Many arbitrators will view the last-chance agreement between the employer, union, and employee as a binding modification to the collective bargaining agreement. This being the case, the arbitrator would be very limited to review the last-chance agreement unless

- The last-chance agreement was entered into unknowingly by the employee
- The language was ambiguous and poorly defined
- The agreement was inconsistently implemented/processed
- The agreement conflicted with state law/regulations
- The last action was not part of the last-chance agreement[20]

Under the due process, public sector employees are required to have a name-clearing hearing, if requested. A public sector employee is entitled to a name-clearing hearing because of the liberty interest protected by the due-process clause of the 14th Amendment. In the *Loudermill* case, the U.S. Supreme Court recognized that a right to a name-clearing hearing before discharge or imposition of an unpaid suspension for employees covered by a just-cause protection. The employee need only establish a likelihood of the dissemination of stigmatizing material to prove the need for a name-clearing hearing; that is, the employee does not have to show that an actual stigmatizing action has occurred, only the likelihood of one. Public sector fire departments have an obligation to offer due process to employees who have a property or liberty interest in employment. Private sector fire departments are generally not required to provide a due process for employees.[21] A probationary firefighter may also be entitled to a postdischarge name-clearing hearing, if he or she can establish a likelihood of the dissemination of stigmatizing action when the employee seeks expungement of stigmatizing material.[22]

One example would be a firefighter who has been wrongfully accused of sexual harassment. The stigmatizing action would comprise comments made by the municipality in the media regarding the allegations or when the employer places those comments in the accused's personnel file. In the fire service, it would not take very long for such information to be disseminated across the news media. Yet, the accused has no way of correcting the false information once it has been disseminated. The municipality must balance the rights of the accuser and of the accused. Each may have a claim against the municipality if not handled in the correct manner. Each party must be offered adequate remedies when a sexual harassment claim is filed with the municipality. When a public sector employee is wrongfully accused of sexual harassment, the employee may have several options to file claims against the municipality (defamation, intentional infliction of emotional distress, negligent infliction of emotional distress). Granted these claims are not so easy to win against a municipality; still, the accused has a constitutional due-process claim. As Chiaman Wang has said, "although name-clearing hearings may provide the wrongfully accused an opportunity to restore his professional reputation, this hearing is not afforded to all individuals wrongfully accused of sexual harassment."[23] The responsibility rests in the hands of the accused to request a name-clearing hearing; by not requesting a name-clearing hearing, the accused is waiving his/her rights to a name-clearing hearing.[24]

Under the umbrella of due process, firefighters have several options for seeking relief from actions taken against them. The firefighter may have the option to seek relief from pursuing a grievance with the union for an alleged collective bargaining agreement violation. Some firefighters may be provided with civil service rules that, if violated, allow the firefighter to file a complaint with the state civil service commission. Depending on state statute, the firefighter may be able to file suit in state court for relief. If the firefighter is claiming a violation of a federal statute or a provision of the U.S. Constitution, then he/she may be able to file in state or federal court. As identified, the firefighter may have several options to seek relief; firefighters who believe their rights have been violated should contact an attorney who specializes in this area. As a result of some time limitations with filing complaints, it is best to seek advice immediately to ensure that you do not release any of your protected rights based on time limits for filing complaints.

Final Thoughts

Regardless of a disciplinary action or a simple disagreement, employees want to know that they have a process to air their complaints/concerns and that the process will be effective and fair. The four tenets of discipline are

- Clearly written/established rules/regulations
- Knowledge of the rules/regulations
- Knowledge of the penalties for violating the rules/regulations
- A process in place to file a complaint that will have an appeals process involved

A discipline process that has only one level of appeal will generally be viewed in a negative light. When allowed to appeal prior decisions to the next level/process, employees are provided with a different form in which to present their case.

Employees who have been wrongfully accused of a claim are entitled to a name-clearing hearing under the 14th Amendment. Further, municipal employees may receive additional rights under the state constitution, as municipalities must provide the same level of due process or more protection than the 14th Amendment. The public sector fire department has an obligation to offer due process to employees who have a property or liberty interest in employment. Private sector fire departments are generally not required to provide a due process for employees. At the end of the day, the objective of a disciplinary program is to create and maintain a productive, responsive, and safe fire department.

Key Terms

- 14th Amendment
- At-will-employee
- Cause
- Culpable behavior
- Discharge
- Disciplinary action
- Due process
- Employee rights
- Equal Employment Opportunity Commission (EEOC)
- Forced leave of absence
- Inventory search
- Just cause
- Last-chance agreement
- Liability
- Name-clearing hearing
- Nonculpable behavior
- Policy update/edit
- Probationary clause
- Separation agreement
- Suspension
- Reprimand
- Termination
- Threat Assessment
- Warning
- Workplace aggression
- Zero-tolerance policy

Exploring Issues on the Web

Last Chance Agreement Arbitration Award: http://www.fpelra.org/Arb_Awards/Arb%20Decision%20St%20Pete.pdf (Arbitration award over the merit of a last-chance agreement. This may be a good resource for students to read and discuss why they agree or disagree.)

References

1. Aitchison, W. 2010. *The Rights of Firefighters*. 4th ed. Portland, Ore.: LRIS Publications. P. 62.
2. Cole, N. 2007. Consistency in employee discipline: an empirical exploration. *Personnel Review*. 37 (1): 109–17.
3. Dessler, G. 2008. *Human Resource Management*. 11th ed. Upper Saddle River, N.J.: Pearson Prentice Hall. P. 564.
4. Aitchison 2010, p. 310.
5. Bohlander, G., and S. Snell. 2010. *Managing Human Resources*. 15th ed. Mason, Ohio: South-Western Cengage Learning.
6. Cayer, N. J. 2004. Public Personnel Administration. 4th ed. Belmont, Calif.: Wadsworth/Thomson Learning. P. 94.
7. Cadrain, D. 2007. Don't let dormant policies lie. *HR Magazine*. 52 (7): 74.
8. Ibid.
9. Dessler 2008, p. 565.
10. Outten, W., R. Rabin, and L. Lipman. 1994. *The Rights of Employees and Union Members*. Carbondale, Ill.: Southern Illinois University Press. P. 19.
11. U.S. Equal Employment Opportunity Commission. 2010. Job bias charges approach record high in fiscal year 2009. *EEOC Reports*. http://www1.eeoc.gov/eeoc/newsroom/release/1-6-10.cfm?renderforprint=1 (accessed October 5, 2010).
12. Fried, B. J., and M. D. Fottler. 2008. *Human Resources in Healthcare: Managing for Success*. 3rd ed. Arlington, Va.: Health Administration Press. P. 119.
13. Van Bogaert, D., and A. Gross-Schaefer. 2005. Terminating the employee-employer relationship: Ethical and legal challenges. *Employee Relations Law Journal*. 31 (1): 49–66.
14. Fried and Fottler 2008, 133.
15. Cihon, P., and J. Castagnera. 2008. *Employment and Labor Law*. 6th ed. Mason, Ohio: South-Western Cengage Learning. P. 3.
16. Ibid., p. 8
17. Fried and Fottler 2008, 134.
18. U.S. Const. amend. XIV, Section 1.
19. Varone, J. C. 2007. *Legal Considerations for Fire and Emergency Services*. Clifton Park: N.Y.: Thomson Delmar Learning. P. 311.
20. Arizona Industrial Relations Association. 2001. Does just cause apply to last chance agreements? *Exchange*. Issue 7. http://www.lera.uiuc.edu/chapters/arizona/issue7.pdf (accessed October 8, 2010).
21. Varone 2007, p. 318.
22. *Cox v. Boxer*, 359 F.3d 1105 (9th Cir. [2004]).
23. Wang, C. 2010. Name clearing hearing: How this "remedy" fails to safeguard the procedural due process rights of public employees accused of sexual harassment. *Georgia State University Law Review*. Summer. http://www.lexisnexis.com.ezproxy.snhu.edu/hottopics/lnacademic/?verb=sr&csi=165268 (accessed October 8, 2010)
24. Ibid.

HUMAN RESOURCE MANAGEMENT 8

Learning Objectives

After reading this chapter, you should be able to:

- Describe the importance of managing human resources.
- List the six areas of the management process.
- Describe the personality traits of firefighters.
- Describe the human resource triad.

Personnel management requires that everyone in the department—rather than just one or two groups of people—be involved in the process. How an organization deals with its personnel will often determine its success. A fire department that encourages its personnel to feel as a truly a critical element of the makeup of the department will yield great results. Human resource management can play a significant role in retaining personnel. One researcher has placed the cost of personnel replacement cost at 70%–200% of the employee's annual salary, in addition to the institution's knowledge loss.[1] Given the significant cost of personnel replacement, it is advantageous for the fire department to have a solid human resource program.

Depending on the size and makeup of your department, the function of human resource management may fall on the fire chief or on the city/town human resource director. In some cases, no one serves officially in the position of human resource manager; rather, the tasks are incorporated into a number of positions within the fire department. Overlooking the function of human resource management could be a costly mistake—leading to low morale, high turnover, and substandard training among other problems.

This chapter will provide a basic overview on how human resource management can contribute to the achievement of the labor-management relationship and improve personnel management. Some fire departments in the United States have begun to see the importance of utilizing the talents of their personnel as well as the ever-changing work and social environment. The functions comprising human resource management are as follows:

- Planning
- Organizing
- Staffing
- Controlling
- Leading
- Directing

Much has been written on the topic of human resource management. This chapter focuses on how certain aspects of human resource management can have an impact on the labor relations process and on personnel management. As a result of space limitations, the topic of human resource management has been distilled into the areas most important to labor relations. By no means is this intended to underestimate any other areas of human resource management. As you read, please envision how the recommendations, solutions, theories, and ideas presented here can be applied to your individual department and your personnel management system.

Why Human Resources?

It has been said that organizations compete through people. As a fire chief, your intent should be to recruit and hire the best personnel. This will provide your department with the most highly skilled firefighters in the area while reducing the training cost to your department. The responsibility to hire the best is not just the fire chief's responsibility. The individual members of the fire department have a responsibility to want only the best standing next to them. When the entire department takes an active role in the recruitment process, this increases morale within the organization and reduces the cost of training/retention—savings that can now be applied to other areas within the department.

In some cases, having a group of specialized trained firefighters can provide a bragging privileges or serve as a prestige issue for the department. Highly talented firefighters can also serve as a model for the department and possibly the area. Face it: you have all listened to others who claim to have, for example, the first and the best dive team in the area or the best high-angle rescue team. Yes, firefighters are a highly competitive group of people. The following sections examine some of the most critical elements for recruiting, hiring, and retaining firefighters. In the simplest terms, human resource management is about the utilization of individuals to achieve organizational objectives and goals.

So, what role does human resource management play in utilizing individuals to achieve the fire department goals and objectives? Generally, human resource management consists of myriad functions. Some human resource management systems will conduct all the functions within the organization, others may have limited roles within the organization, and still others will contract systems to outside vendors. As for the fire department, larger departments may have their own human resource manager, while other departments may rely on the municipal human resource manager. In smaller departments, this function often falls onto the fire chief. Therefore, it is essential for the fire service to have a basic understanding of the role human resource management can play and how it can impact the labor relations.

Labor unions are concerned with rules governing the employee and the organization and how they are enforced. Human resource management may be charged with many issues affecting the individual employee and the organization (e.g., wages, benefits, laws, regulations, and insurance) that have an impact on the labor relations process. Human resource management can also play a role in employee well-being and performance at work. Human resource management does not always have to be a business-oriented performance outcomes system, which is one-sided.[2] Personnel management is intertwined with human resource management.

Human resource management functions may consist of staffing, human resource development, compensation, safety and health, and employee and labor relations. In The fire department as an organization is affected by various areas, some of the which are overlapping and others of which stand alone (table 8–1). This chapter focuses on the staffing, human resource development, compensation, safety/health, and employee and labor relations. Labor relations is just one small piece of a larger puzzle encompassing the operation of the fire department.

In this chapter, the following items are considered to have a direct impact on the department personnel:

- Productivity
- Morale
- Employee satisfaction

The first-line officers in the fire department may have the greatest impact on the line firefighter when it comes to creating a positive or a negative work environment. This does not remove the fire chief's share of responsibility for creating a positive work environment for all employees; the positive or negative environment will trickle down through the department. Thus, the culture of the department will take shape over time and cannot easily be transformed (or repaired) overnight. As a new fire officer, you may be confronted with changing this culture. As a fire officer you should set the tone, and the best way to achieve this is always to promote doing that the right thing (fig. 8–1).

Fig. 8–1. Fire department environment

Human Resource Management Administration

Experts in the field of management have identified the basic functions that mangers perform, called the *management process*. The management process is utilized by fire department administrators; municipal, state, and federal government; and the private sector. Many labor unions today are run like a business, and in larger unions, the management process can be utilized in many of the same ways. So if you are a union official, you may want to see how your union can adapt the following management process to ensure a better-run organization. If you are a fire service administrator, you should evaluate how well your fire department performs these critical points. Volunteer fire companies may also utilize these items in providing a better-run organization. Human resource management is also a critical piece when it comes to blending the five elements of the fire service together (fig. 8–2):

- Customers
- Employee
- Technology
- Capital
- Time

To ensure that the above areas are properly addressed with adequate resources, managers and administrators turn to the formal management process.

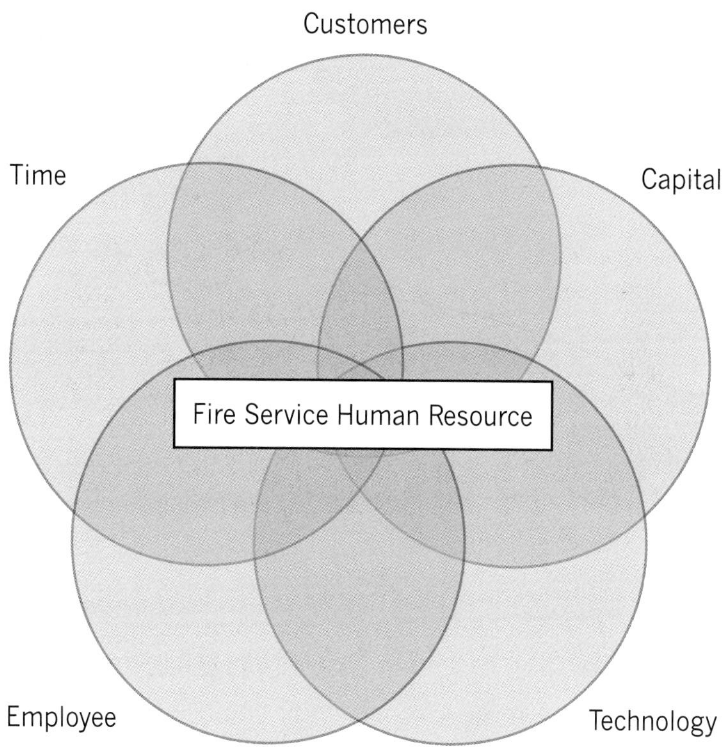

Fig. 8–2. Fire service human resource management

Fire service administrators will often admit that the employees are the department's most critical resource. What is an organization without people? Often, the general public will judge a fire department by its equipment and/or station. This estimation could not be further from the truth—the people make the department. The foundation of the fire department to operate rests on the willingness, dedication, commitment, and ability of its firefighters to make things happen. Human resource management plays a role in ensuring that human resource needs are met and that you have an effective organization.

The six steps in the management process will be discussed in detail in this section (fig. 8–3).

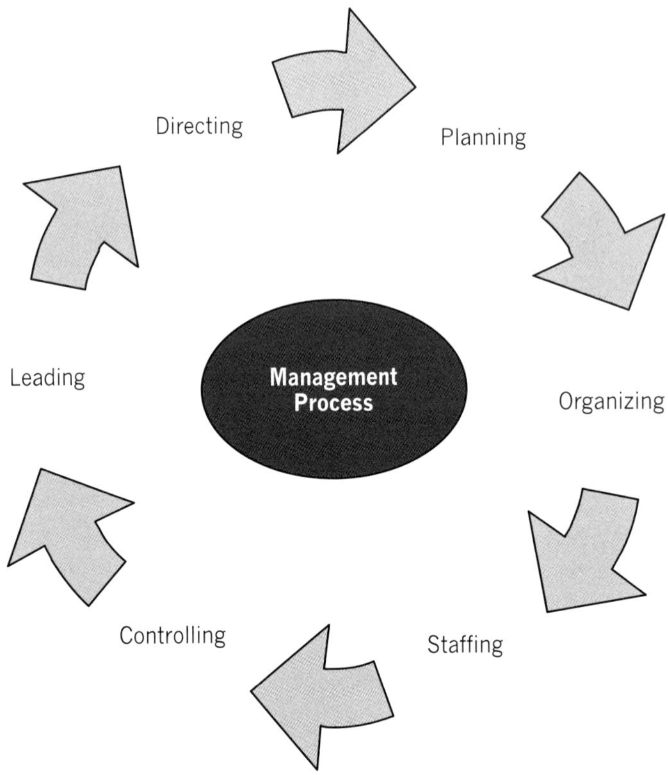

Fig. 8–3. Six steps of the management process

Planning

Every program needs a plan. Without a plan the program will not have direction. The planning process must also be coordinated with program evaluation.

People often underestimate or overlook program evaluation. The evaluation process examines those changes that have taken place and scrutinizes whether they are the intended changes projected by the project team. Hence, the evaluation process allows for readjustments in the program.

Human resource management should be a segment of the fire department's strategic planning process. In larger fire departments, these areas are discrete divisions operating independently; however, the planning process allows the human resource manager to align and realign the fire department's policies, procedures, laws, regulations, collective bargaining agreement, and organizational change, many of which are current issues facing the department.

Another function of human resource management planning is developing a plan for the next 5–10 years. Such long-term human resource management planning entails forecasting the department's personnel needs, equipment needs, facility needs, and any internal or external factors affecting the department. The long-range plan should be reevaluated each year to ensure that the fire department is moving in the right direction and to make changes as needed.

Making a change is itself a process that requires planning. Fire service administrators will implement change because at the time it seemed like it was needed. Keep in mind that, just because you made a decision and change things, nothing prevents you from changing your mind later, as additional information becomes available. There should be no shame in admitting that, although you made the best decision with the information that you had at the time, new information has come to light that necessitates adjusting your decision.

It is impossible to predict the exact needs of the department in 10 years. For example, the fire department management team may develop a plan that calls for increasing the number of firefighters by 45% in 10 years; however, in years 6, 7, and 8, the town sees a significant outgrowth in the most northerly section of town that increases the number of homes and businesses significantly, beyond what had been forecast. During this time frame, at the annual reevaluation planning process, proper adjustments should be made to accommodate the current changes facing the fire department.

Another factor to consider in staffing is the projected number of firefighters who will soon be retiring. Over the next couple of years, this will be evident across the fire service as the baby boomer firefighters retire. Wayne Mondy, author of *Human Resource Management*, states that "changing conditions could affect the entire organization, thereby requiring extensive modification of forecasts. . . . Planning, in general, enables managers to anticipate and prepare for changing conditions, and [human resource] planning in particular allows flexibility in the area of human resource management."[3] The number of potential members facing retirement in the coming year(s) is often great in fire departments that had a significant increase in personnel 30 to 35 years ago, as a group of employees hired together will generally retire around the same time, resulting in a loss of experienced personnel for a fire department.

All too often, fire service administrators are reactive to the changes in the community. If you currently have a significant need for additional personnel, then chances are you are too late. The reality is that it will take 12–24 months to sell your plan to the municipal officials, secure funding, advertise, hire, and properly train a firefighter. And this is assuming the funding is approved in the first year you ask for it; in many cases, it may take a second year to gain funding and local support for your plan. Regardless, without proper planning, you and your department will be foundering—trying to bail out a sinking ship with nothing more than a teacup.

The planning process should be widely used within your local fire department. Without proper planning, the fire department will soon find itself outdated and lacking the resources to accomplish the mission-critical tasks. The planning process, when used properly, can establish the levels of service desired, efficient and effective use of resources, and implement a management system.[4]

Organizing

Once the planning process has been completed, the next step is organizing, in which it is determined how the project will be accomplished. Organizing is a sound management practice to ensure that the project has ample resources. The manager will group items needed to areas in an effort to carry out the project's objectives.

During the organizing process, the project manager will enumerate the various tasked to be performed and determine who will be responsible for each task. In some cases, the tasks may be performed by a firefighter; many times, though, a shift, a team, or even a division within the fire department will be assigned the tasks, and organizing these teams is essential. While teamwork is no stranger to the fire service, the administrator sets the tone for creating an environment that is conducive to teamwork. The project manager must also assign the authority to perform the task for the project. Think of this step as establishing an *activity-authority network* within the fire department.[5]

Local, state, and federal government use more or less the same methods for organizing. In recent years, the organizing methods used in the public sector have received criticism.[6–8] The fire service needs to rethink how we organize our projects, as the old method may no longer work for the department and the community. A fire department may have to alternate between various styles of organizing, depending on the task at hand. Regardless of the task, organizing is part of the fire service, day in and day out.

Staffing

The staffing process involves recruiting personnel who are the best qualified for the open position. This may involve selection of both line firefighters and civilian employees of the department. In a union, the salary and benefits plans are generally known in advance, so the new employee is aware of what the entry salary will be. This eliminates the process of bargaining over individual wages and benefits. The exception to this is when the collective bargaining agreement has provisions for a new employee to jump steps on the basis of prior experience and training, which is often agreed on beforehand with the labor union. Some communities are in the process of examining this process in an effort to attract seasoned firefighters. What the new employee who jumps steps would generally not be granted is his/her seniority from a prior fire department; the only exception to this would be a firefighter transferring into the police department within the same community, who might be allowed maintain all seniority and benefits within the new department. This is contingent on the language contained in the collective bargaining agreement.

The human resource manager's job does not end when the firefighter is hired. The human resource manager is responsible for *on-boarding* the new employee—providing all the insurance documents, department policies, retirement information, medical information, and any other information pertaining to new employees.

Hiring the wrong personnel can be a costly mistake for the fire department, the labor union, and the entire organization. Unproductive and troublesome personnel will challenge the labor-management relationship and may lead to an increase in grievances stemming from progressive discipline issues, and if the employee is terminated, the local union may be faced with defending the member. Consequently, it is safe to say that labor and management both want to hire the best person for the job. Moreover, involving the labor organization in the hiring process may provide the fire chief with a different viewpoint and could yield a better candidate, whereas having only one level of management may not be the best approach to hiring a line firefighter.

Today's service administrators need to look beyond the task of fighting fires when it comes to recruitment. The skills and ability of the firefighters you hire now should be able to drive the department into the future. Modern firefighters must be able to adapt and learn new skills—dealing with chemicals, medical skills, building and plan design, budgets, procurement, project management, and public speaking, just to name a few. It comes down to what your department needs. It would be unfair to develop a list of what every firefighter should have; while you may be able to list the base set of skills and abilities, much more needs to be determined at the local level.

The community may also be faced with liability, under the allegation of negligent hiring, if employees cause harm to others. The theory is that the employer has a responsibility to provide the employees with a safe work environment, free of dangerous equipment, for example; a dangerous employee could be viewed in the same light as a dangerous piece of machinery. Legal precedence has evolved such that negligent hiring now extends to injury or harm to a customer. The negligent hiring theory forces the fire chief and the community to hire the "safest employees." In *Garcia v. Duffy*, the courts said, "negligent hiring occurs when, prior to the time the employee is actually hired, the employer knew or should have known of the employee's unfitness, and the issue of liability primarily focuses upon the adequacy of the employer's pre-employment investigation into the employee's background."[9]

The human resource manager should also be involved with the fire department training officer to ensure that all state-mandated employee training requirements are met, including but not limited to sexual harassment prevention and education, sensitivity training, awareness of employee assistance programs, and retirement information. The training process could be broken into two programs:

- New employee education
- Career development

Another area often overlooked in the fire service is *employee appraisal systems*. Management and the labor organizations are often at odds about this topic unless each side has been involved in the development of the system. The human resource manager can assist the fire officer in appraising a firefighter's performance against previously approved standards; however, the real challenge is for labor and management to agree on the approved standards before the program is implemented. After conducting appraisals, fire officers need to counsel the firefighters on how to improve their performance. Fire officers and human resource management should be involved with career counseling of all personnel, from preparing individuals for promotion to providing all personnel opportunities for development.[10] The process should not be about only finding fault or deficiency; rather, the employee appraisal system should advance the personnel through professional growth and development. At the same time, the process allows the fire department to align the employee's goals and development activities with the department's priorities.

Forecasting the number of personnel, skill level, and location of firefighters is critical to sound human resource management. Two well-used practices for forecasting future staffing needs are *zero-based forecasting* and *bottom-up forecasting*. Zero-based forecasting uses the current fire department staffing as a starting point for determining future needs and is both the most common and the easiest approach used in the fire service. Zero-based forecasting is based on the same principle as zero-based budgeting.

The bottom-up approach requires each division in the fire department, to develop an aggregate forecast of firefighters needed. Who better to decide whether more or fewer employees are needed than the manager of a specialized unit? This process works best in large fire departments and departments with specialized units. Although the process is labor intensive and time consuming, bottom-up forecasting provides a form of checks and balances, each year and over the projected time line, to make sure that staffing is being utilized to its fullest. This approach may result in staffing readjustment to better meet the current needs of the department. Too many times, I have seen fire departments request more staffing when they lack data to back their request; instead, they might use the call volume to support the argument for additional staff. Bottom-up forecasting is an interactive approach that involves the different levels within the fire department. The negative side of this approach is that each manager or fire officer is forced to justify anticipated staffing needs.

Fire departments that are nonunion should review current hiring policies and agreements, to limit potential liability by use of an *expressed employee contract*. An expressed employment contract defines the purpose of the employment relationship and can be enforced in the court of law. The following are three significant issues that the municipality should be concerned with under an expressed employee contract:

- *Length of service*. Do your personnel programs and practices imply that the employee will be granted an unlimited term of employment?
- *Employee dismissal process*. How will the employment relationship be terminated and under what conditions?
- *Breach of contract*. It is recommended that the municipality clearly limit the remedies that the employee can seek for a breach of contract—for example, allowing arbitration to settle any disputes with a final and binding decision. In this example, the two parties could not utilize the judicial process.

The nonunion fire department may want to consider the benefits and restriction of offering individual employee contracts to new employees.

The second item that a nonunion fire department should consider when reviewing hiring guidelines, policies, and employee handbooks is the *implied employee contract*. In this case, the employment contract is not written but is mutually agreed on at the time of hire. The municipal government and fire department handbook, policies, and procedures may be written in a way that infers an implied employee contract. The courts will also look at the behavior of the two parties as they pertain to the law. Even in states with at-will employment, an implied employee contract can be considered and upheld by the courts. It is recommended that the legal department for the municipality review this process for limiting litigation due to claims of wrongful discharge. (The issue of termination of employment is discussed in greater detail in chapter 7.)

The economic downturn from 2009 to 2011 has led to fire station closings, personnel reduction, and budget cuts of the likes that most firefighters have not seen in the past 20–30-plus years. The human resource department will be involved in the process of downsizing agencies and personnel. Carlos Migoya, city manager of Miami, recently imposed $76.9 million in salary, health insurance, and pension cuts because Miami was faced with a deficit that resulted in reduction of the city's credit rating; consequently, the city manager declared an urgency to deal with the cuts. Some states have laws that allow or require municipalities to renegotiate collective bargaining agreements during significant fiscal distress. This issue is not an isolated issue pertaining to Miami alone; rather, it is a common fiscal issues faced by municipal leaders and public sector unions, with little relief in sight.[11] Human resource departments will be required to aid in the renegotiation process of the collective bargaining agreements, to deal with personnel reductions, and to project pension plan assets against municipal obligations down the road. How significant these reductions in services will be and how long they will last is yet to be determined. How reductions in services and personnel are handled now will have a significant impact on labor relations and personnel management in the future.

Staffing is a large and detailed process, a full discussion of which is beyond the scope of this book. However, because staffing is an important segment of labor relations, if this area is overlooked or rushed, then poor decisions are made and the labor relations process may be challenged.

Controlling

Management will utilize control in an effort to make sure that resources are utilized in an effective and efficient manner in obtaining the organizational objectives. Management control has been defined by Langfield-Smith as follows: "Management control systems provide a means for gaining cooperation among collectives of individuals or organizational units who may share only partially congruent objectives, and channeling those efforts towards a specified set of organizational goals."[12] In simple terms, we ask the following:

- Is the task being accomplished?
- What is not being accomplished?
- What progress has taken place?
- What resources are still needed?
- What changes/deviations have taken place since the beginning of the plan?

The union executive board and elected officers may establish a plan, goals, and objectives for the following year. At established benchmarks, the board will review the plan and the work that has been accomplished to date. During the controlling process, the executive board can take any corrective actions should it see that the objectives are not being met and may revise the plan and objectives if warranted. This would not be possible if it did not have a plan, goals, and objectives. The very same process is often used in many fire departments to ensure that resources are being used efficiently and effectively.

If you are a fire chief or a company officer, you will need to exercise control with your personnel. If personnel are allowed to do what they want, when they want, then your department will not function properly, with the end result conceivably being property loss or loss of life. Personnel management is central to the effective and efficient use of human resources. A fire service leader will often use different control methods for different incidents. For example, the fire chief will not use the same personnel management system on the fire ground as at the fire station when dealing with a troubled employee. Different incidents require different styles of management.

Personnel management also involves the advertisement for, hiring, promotion, transfer, and demotion of employees under certain laws, regulations, mandates, and contractual obligations spelled out in the collective bargaining agreement. When a fire department oversteps its legal authority, the case may be reversed in the court of law or by means of alternative dispute resolution; this can be costly and a waste of time and department resources. Labor organizations are often concerned with how employees are hired, promoted, transferred, demoted, and/or terminated, and in many cases, these issues are addressed in the collective bargaining agreement, along with a detailed process outlining the accomplishment of each. When the collective bargaining agreement is silent on such personnel issues, look to your department policies and procedures and/or the city/town employee handbook.

On a smaller scale, the company officer or the shift commander may have the responsibility for scheduling vacations, approving overtime positions, and preparing payroll. These also require controls that ensure that the organization is running effectively and efficiently. The company officer or shift commander who is scheduling vacations or approving overtime must also ensure that he or she is working within the confines of the collective bargaining agreement and any of the city or town's policies and procedures.

Another area that involves control in the department is the organization's need to adapt to a changing legal system, mandates and regulation crafted by others, budget control and limitations, and the needs of the community. Much of what we do within the walls of the fire station and how we perform our jobs in the street is dictated by laws and regulations. It is the responsibility of every firefighter, fire service administrator, and union official to stay abreast of changing laws, recently issued rulings in court cases, and pending legislation that affects the fire service.

In the perfect world, the fire department has all of its wants, needs, and a never-ending pile of money, in return for which the community is provided with first-rate service. Unfortunately, this is not the case in the real world: controlling cost is a fire service administrator's main concern. The fire service administrator can make only so many legitimate reductions in the area of materials and goods before personnel costs, which generally make up a significant part of any fire department budget, must be reduced; hence, labor relations can be costly depending on how personnel management is carried out.

Leading

Grover Starling has defined the leader's role as a "relationship with subordinates, including motivation, communication, encouragement, and influence."[13] One challenge the union faces is when the union president's word is different in public than in private. The same holds true for the fire department; a fire chief whose words are different in public than in private runs the risk of hurting the organization and losing the ability to lead. A contradictory voice can ruin the career of a leader of any organization. At this point, I am sure you are thinking of your own example.

Day and Lord have found that up to 45% of an organization's success can be directly linked to the leadership of the organization.[14] In these terms, a fire chief with no leadership ability significantly reduces the success of the department. The fire chief may be a good person, but is he or she cut out for the job? In smaller communities, you will often see community leaders appointing the most liked person as the head of the fire department. A great firefighter does not necessarily yield a great fire service administrator.

It is time that community leaders start hiring administrators with solid leadership abilities. Gone are the days of having a good person in the corner office. Just because the current fire chief's grandfather was the fire chief does not mean that he will be a good chief. In one department, only after three years did the firefighters find out that the nice guy who was their chief had done nothing to improve the department and, if anything, was not fighting for his firefighters. Five years later, the nice guy was still the chief, and the employees were still without a labor contract. Having no contract for so long and the resultant very low morale stagnated the growth and potential of this department. This was all due to the elected town officials' poor selection of a fire department administrator.

Labor organizations are no different: union members must elect a leader as union president. This is your union, and the success of the union can be found in the leadership. In some locals, the election for the office of president can be heated, with members fighting until the last ballot is entered. In other unions, members sit around the annual meeting looking at each other and asking, "Who wants to be president?" Again, up to 45% of the union's success can be linked to the leadership of the organization. Stand up and be counted. Take control of your labor organization. Recruit and elect a leader as president next time around. If you wonder why your labor relations are not working, then maybe it is time for a change in leadership. You have a say in your union, so make sure your voice is heard.

Having a partnership with the union when it comes to human resource management should entail a comprehensive plan. As you have seen, the fire department is involved with and affected by many of the functions of human resource management. Table 8–1 outlines possible considerations when dealing with a labor union and performing human resource management. The list in table 8–1 is not all-inclusive; rather, it is intended to spark your thinking process, in terms of ways in which the organization and the labor union can work together for the betterment of the fire department.

Table 8–1. Human resource management activity

HRM Activity	Industrial relations considerations
Planning	Will the union and/or employee representative be involved in the formulation of HRM plans? Will the plan be subject to revision or modification after negotiation with the union, or the subject of revision after consultation with employees?
Recruitment	Does seniority apply in the case of the position to be filled? Are there provisions contained in the collective bargaining agreement that apply to the position being filled? Are there provisions contained in the collective bargaining agreement that apply to internal versus external recruitment?
Selection	Will preference be given to members of the union, or will this be a non-issue? Will the union and/or employee representative be on the selection committee? Does the collective bargaining agreement have provisions pertaining to selection?
Performance appraisal	Does the collective bargaining agreement have provisions pertaining to appraisal process? Is there a provision in the collective bargaining agreement on disciplinary procedures for non-performance? What provision does the employee have available to appeal the appraisal if viewed as unfair? Are personnel policies contained in a handbook or the labor-management agreement? Does the labor-management agreement contain provisions dispute resolution? Should the union and/or employee representative be part of the appraisal process?
Training and Development	Does the union need to be consulted on training programs? Are there provisions contained in the management-union agreement on skill development and training programs? Does the collective bargaining agreement contain provisions on outside training leave? What nationally recognized training standards will be utilized?
Health & Safety	Will the union and/or employee representative be involved in the formulation of a health & safety plan? What standards will be used? Will a labor-management safety committee need to be formed? Is the health and safety plan operating effectively from a labor relations and legal point of view?

Table 8–1. Cont.

HRM Activity	Industrial relations considerations
Compensation	Are there provisions contained in the labor-management agreement pertaining to wage issues? Will the union be involved in any job evaluation and skills auditing exercises? What are the flexible fringe benefits contained in the labor agreement and the organization? Are all the compensation programs within the legal limits? Will the union be involved in any compensation audits?
Civil Rights and Equal Opportunity	Will the union and/or employee representative be consulted on equal employment opportunity policies or other "work programs"? Will the union and/or employee representative be involved in the preparation of "work program" reports? Does the labor-management agreement contain provisions on ethical standards? Does the organization meet the requirement for civil rights, EEOC, and diversity requirements? Has an audit been completed? Does the organization meet the requirements of any and all employment laws? Has an audit been completed?

Source: (Adapted from) Abbott (2007, p. 68-69)

Directing

The area of directing involves the fire service administrator making decisions, issuing policies/directives/guidelines, and directing the work that needs to be performed. If the management system falls short in the area of directing, it reduces the success of planning, organizing, and staffing. In the simplest terms, directing involves the process of providing guidance to people on how to do the job; for example, the shift officer may provide guidance and inspiration for the line firefighter, while the fire chief may delegate his authority to investigate fires to the captain in charge of fire prevention. The process of directing is an active process within the organization and is used on projects and during emergency incidents. Directing will often involve the following areas:

- Communication
- Supervision
- Motivation

Strategic Human Resource Management

Many fire departments are developing strategic plans for the department in an effort to maximize the present and future efficiency of the fire department. To take the planning process one step further, we must look at strategic human resource management. Strategic human resource management is nothing more than a part of the entire strategic planning process for the fire department; in simple terms, strategic human resource management weighs internal strengths and weaknesses against external opportunities and threats, to provide a competitive advantage. When teaching strategic planning, I ask my students: "Where are we now as a fire department; where do we want to go; and how should we get there?" Hence, the strategic human resource management plan is nothing more than a road map for human capital—developing a strategy and turning it into an action.

Human resources strategies will often look at the fire department productivity and performance improvement areas. The strategic plan demonstrates that employees play an active role in the fire department's productivity and performance efforts. The human resource management office must be part of the design, implementation, execution, and follow-up on the fire department's strategic plan. Essentially, the human resource strategic plan is just one small piece in the overall fire department strategic plan.

Firefighters' Personalities

Whether you are a supervisor or are involved with a labor organization, you should have a basic idea of the impact that firefighters' personalities can have on the service that they provide, as well as on their ability to interact with people. Knowing about personality traits of firefighters may help predict how they may respond when dealing with personnel issues.

Spend any time in the fire service and you will quickly realize that it takes a very special person to do the job. Alternatively, some would say that that the fire service culture affects how the firefighter behaves and changes the firefighter's personality. What makes up a firefighter's personality is what drives this person. Research by Peters and Scott has shown that personality traits of firefighters are different than those in the general population.[15] In addition, the same characteristics that drive firefighters to do their job can be the very characteristics that destroy a career (fig. 8–4).

Tinkerer

Firefighters are crafty people, always seeking a better way of doing things. Status quo is not in the vocabulary of the firefighter. Think of a task that a firefighter must perform and ask several firefighters how they think they can do the task more quickly, easily, or better—in short order, you may have an answer. If not, wait a minute, and they will come up with something. The time clock often does not reflect the true beginning and end of the work day for firefighters; they are always thinking about the job and how they can be better at it. As a line manager, it is best to keep your firefighters motivated and active. When firefighters have too much down time, they will find ways to entertain themselves—and not always in the best manner.[16]

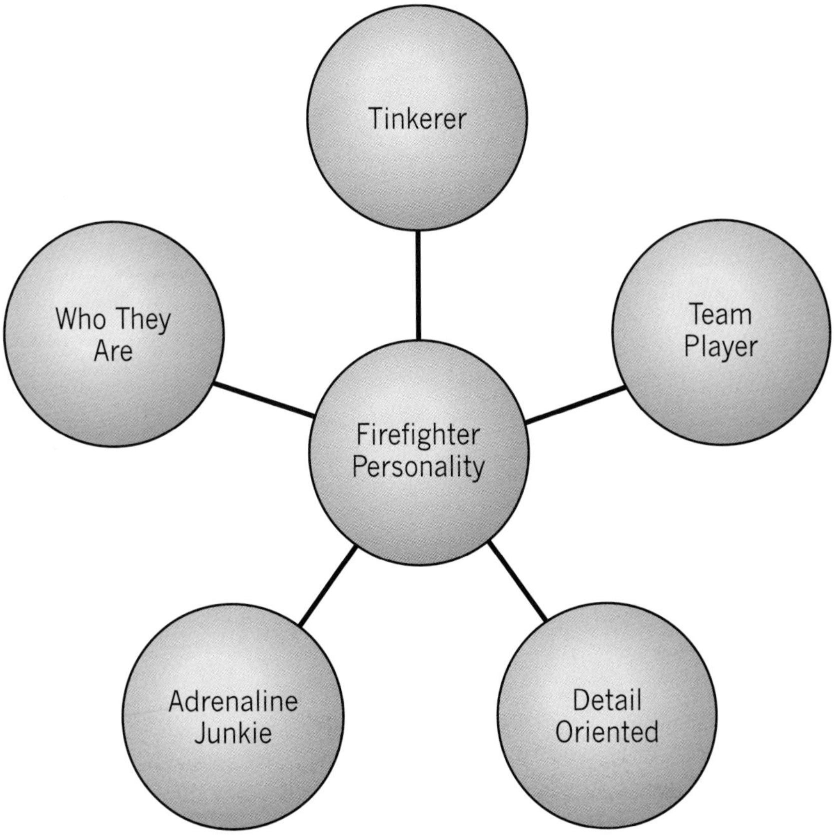

Fig. 8–4. Firefighter personality traits

Team player

A firefighter is taught from the very first day on the job that the fire service is based on a team approach. Work assignments are often given to teams in the station and on the emergency scene; therefore, firefighters function in a group environment. Living and working in such a close-knit community shapes firefighters, much as parents, school, and peers shape adolescents.[17] It is also not uncommon to find that firefighters will seek team-based activities outside of work. The negotiating team is no different; research and data collection are not done solely by the chief negotiations office or the union president.

As a result of living and working in this group environment, over time, the team will develop team values, form team cohesion, and identify team members' strengths and weakness. When a fire chief is promoted from within the organization—or when any member is promoted within the department—the group generally has an understanding of what that individual's strengths and weakness are. If the former union president—now the current fire chief—is sitting across the negotiating table, then the current union president may have knowledge of the fire chief's weakness and strengths; if the negotiation process becomes adversarial, it is likely that the union would attempt to exploit the fire chief's known weakness.

When an outside chief is hired, it generally takes the local firefighters longer to determine the chief's strengths and weakness. Hence, a new chief has a honeymoon period during which the community, the department, and the labor organization will determine what the new chief brings to the table. Over the next couple of years, the firefighters will buy into the new direction and leadership of the department, or they will resist that new leadership style. During the process, be it good or bad, the firefighters will utilize the team approach. Some argue that one approach is to *divide and conquer* the team when it comes to local politics and implementing change.

My point is that the team approach can be simultaneously healthy and unhealthy for the fire service. As a firefighter, you should have some basic understanding of this and respect the team environment while realizing that is can nevertheless be used against you. As a student of labor relations, you recognize that it is best to know as much as possible about the other side—especially the people you will be negotiating against. Thus, the firefighter has an *interaction-orientation* role, by participating and feeling that the group needs the individual firefighter. The role that a firefighter plays in the group setting may be one of compromiser, harmonizer, encourager, or even follower.

Detail oriented

Take a look around any fire station and you will see just how detailed firefighters are. Some argue that firefighters are obsessive-compulsive in how much attention they pay to details. Firefighters are trained and drilled repeatedly to ensure proficiency in a skill set. This attention to detail is what drives the firefighter to tinker with things to make them better. During negotiations sessions, you may find that the firefighters are detailed in research and data collection; the same can be said for filing grievances, which are often detailed and supported with data.

The firefighter assumes a *task-orientation* role by seeking additional information or clarification from the supervisor or the team. Some firefighters may take on the role of coordinator in an effort to bring a group or idea together. Other firefighters may take on the energizer position that provides the group with energy to complete the task by pushing them to take it to the next level.

Adrenaline junkie

Firefighters are often driven by the natural high they receive from the flood of adrenaline dumped in the body with the notification of a fire or a rescue operation. They enjoy and, at times, thrive on the adrenaline rush of the job. Firefighters will often seek out off-duty activities that provide the same adrenaline rush. Firefighting is a high-risk occupation; firefighters are risk-takers. Risk is what supplies the adrenaline.

Who they are

Generally the firefighter is motivated and driven by the satisfaction of firefighting, and limited satisfaction is derived by the pay and benefits offered to firefighters.[18] Firefighters have great pride and dedication to the fire service, with a deep respect for its traditions and heritage. Firefighters are identified by the job that they do—hence, firefighting is their identity. This is reflected in the off-duty clothing that firefighters choose (e.g., t-shirts, golf shirts, and sweatshirts with firefighter insignia), and a firefighter's personal vehicle often bears some sort of fire service insignia on it. Union firefighters will often place one or more union stickers on a vehicle and also extended-family vehicles as a way to show that they are part of a large social group of firefighters throughout the United States and Canada.

Fire service administrators and supervisors must also deal with destructive behaviors of firefighters and coworkers. The *self-orientation* role is when the firefighter will go out of his/her way to prevent and/or resist any change in the status quo. A firefighter may withdraw from the group in an effort to show that he or she is not happy or does not support the ideas being presented or the change being discussed. A firefighter may take on the position of the blocker, attempting to block new ideas or change from being considered. The blocker may want to revisit past failed attempts, or may just sit back and debate/argue about a topic that the group has considered a dead issue.

Motivation

Fire service administrators and supervisors are often faced with the question: how do I motivate my people? It is interesting to note that one research project determined that some people can maintain their jobs without being fired by working at only 20%–30% of their ability. Highly motivated employees work at 80%–90% of their ability.[19] This being the case, why is it that the issue of motivation is often overlooked? Implementing an employee motivation program can increase firefighter productivity and satisfaction, which in turn leads to the fire department's achieving its organizational goals and objectives.

In the field of social science, much research has been done to develop theories on motivation, and no matter what study you follow (e.g., Hawthorne studies, Maslow's hierarchy of needs, Douglas McGregor's Theory X and Theory Y, or Chris Argyris's personality and organization theory), the fact remains that employees want to work in a environment where they feel their contribution is appreciated, where they receive praise, where they are provided an opportunity to learn from mistakes so that they can improve, and where they have a clear understanding of what is expected of them.

Change can play a role in motivation. People like to know that things are about to change. This means that administration should involve all the people within the organization in the change process. Provide ample information to the entire organization in a timely manner (e.g., via face-to-face meetings, e-mail communications, and department blog postings). Avoid releasing bits and pieces of information about the possible change, as this leaves the window open for rumors to develop and for people to invent their own version of how the change will take place. If firefighters know that a change is about to take place, understand the need for the change, and are given an opportunity to be involved in the change process, then, as individuals and a group, they are more inclined to accept and embrace the change.

Incremental change entails many small changes going on at the same time, whereas radical change is when the change impacts everyone and everything in the fire department. A radical change should involve extensive planning and careful implementation to be successful. A word of caution: the stress and uncertainty a radical change creates in an organization can cloud the end result. More often than not, fire departments undertake change incrementally.

The issue of change plays right into the issue of labor relations when two parties are negotiating a labor contract. Each side sitting at the bargaining table will need to provide some level of information to the members who will be affected by the changes in the contract. Leaking out small pieces of information about what is being discussed at the bargaining table can be detrimental to the bargaining process. If one of the parties chooses to release information to the people affected by the change, then they should provide this information at a general meeting so that all the people have an opportunity to hear what is being said and how the process is going from the person(s) directly involved. This will decrease the potential for misinformation to be relayed and will help to control rumors.

When any organization looks at motivation, two question often come up: "What is going to be considered 'standard performance'?" and, "How will the organization measure standard performance?" These two questions need to be discussed when any fire department is considering crafting a performance management program. The performance management activities must be aligned with the fire department's goal and objectives, culture, and internal environment. A performance management program will also take into consideration any external factors that will have an impact on the department (e.g., changing laws, political fronts, and the labor market). It is recommended that fire departments ensure that the performance management program does not discriminate against any one group or person. The measure should be used fairly across the department, and it is imperative that practices be job related. Performance management programs should be an ongoing process, not just utilized once a year. The critical element to the program is formal measurement and feedback to the employee with regard to performance. During the feedback session, the supervisor and the employee can identify ways to improve performance.[20] The performance management program involves the process of performance appraisal (which will be discussed further throughout this section).

Motivation is often mentioned in regard to improving performance, and establishing an environment wherein employees are properly motivated to overcome personal or professional challenges is imperative. A fire department without motivation is unlikely to perform well during an incident or even during day-to-day activities. Most humans need some level of motivation to perform well. As motivation is a complex concept, we will merely touch on the topic here.

In the simplest terms, motivation has two elements. The first element involves the process of determining which behaviors to engage in; the second element involves the process of deciding how much effort to expend.[21] In application to the issues covered previously in this chapter, motivation takes on a different light. A firefighter who struggles with alcoholism may need a different motivation than the newly hired firefighter and must constantly reevaluate his or her behavior.

The *expectancy theory* of motivation is simple: The individual chooses the behavior and effort level. The person evaluates the intended behavior and the consequences that will follow the behavior (e.g., recognition and reward).

The employee must be made aware of what acceptable behavior is and what the consequences are for unacceptable behavior. This should be part of the plan for recovery in the event that the employee faces a personal or professional challenge. Fire service administrators and supervisors play a significant role in establishing an environment that fosters positive motivation. When employees are faced with negative motivation by the supervisor or fire chief, there is little or no desire by the employee to change the behavior. Even though the fire chief and supervisors can establish a positive motivation, the motivation still comes from the individual employee as to what behavior they will demonstrate. Having a fire department that supports positive motivation will greatly increase the chance that employees will choose positive behavior and place great effort in the behavior.

A recurring theme in this chapter has been involving all the people at the different levels within the fire service organization. The performance management program is no different. In the following section, you will see how several of the issues discussed thus far can be incorporated into the human resource triad.

Human Resource Triad

When the labor relations process involves human resource management, the end result can be very powerful. This process should be viewed as a transformation of policies and practices in the fire department to produce a better department. Regardless of your final destination, you need a solid direction and a process to obtain that end result. Without directions, you are sure to get lost, as well as waste valuable resources and time. The fire service organization should ensure that the following three players are involved in the policies and practices of human resource management (fig. 8–5):

- Human resource management personnel
- Fire administration
- All employees affected by human resource policies/practices

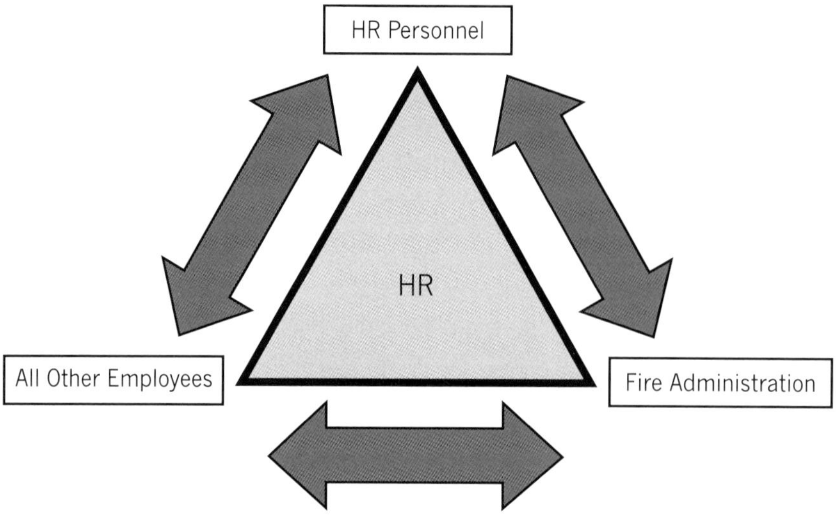

Fig. 8–5. Human resource triad

The human resource management process should be incorporated in to each of the divisions within the fire department. Whether you are talking labor relations or personnel management, the human resource triad should be utilized.[22] Labor-management relations will often be viewed in light of the collective bargaining process; this entails negotiating the contract and also living within the collective bargaining agreement (interpreting, enforcing, and resolving disputes arising from the contract).

Human resource management is generally involved in the fire department process of establishing goals and objectives, developing policies/procedures/guidelines, collectively bargaining wage and benefits, complying with labor laws, and forecasting the department's future needs when it comes to additional firefighters. Labor relations in the fire service involves extensive interactions with the office of human resource management, regardless of the size of your department. The human resource management office for your town can be of great benefit to the fire department. Human resource management has been underutilized in the fire service in the past. Today's trend is to actively engage the human resource management office with personnel issues in the fire department.

Firefighters may think of the human resource management program as merely another bureaucratic office. Often, this line of thinking stems from an entry-level firefighter's first exposure to the human resource management office. When a new employee is hired, there is a lot of paperwork that must be completed before that person can officially start working. This includes establishing a wage step/grade, setting up direct deposit, selecting a health insurance plan, enrolling in a retirement plan, electing state/federal tax deductions, and being issued the department/municipal handbook and emergency notification sheets.

In smaller or midsized fire departments, the fire chief or someone within the fire department serves in the capacity of human resource manager. Few fire departments across the country have the benefit of having a full-time person assigned to a human resource management position; these are often under the administration function of the department. A majority of fire departments across the country are small volunteer and/or call fire departments. In some smaller departments, the task of human resource management is not performed because of a lack of time, lack of knowledge of human resource management skills, or the feeling that the department does not need any human resource management tasks performed because of its size. Nothing could be further from the truth.

The fire service administrators must determine what must be done and by whom. The fire service administrator should be constantly evaluating and reevaluating the specific tasks that firefighters are carrying out. If this area is not evaluated, then the department is missing a significant step in the organizing process. Just because the department has always been run a certain way does not mean that it is the most cost-effective utilization of the departments' resources. However, jobs that historically have been done by the fire chief or the deputy chief may now need to be delegated to subordinates, because of growth within the department or organizational changes. Someone within the department should be coordinating the work of subordinates at various levels to ensure that the work is being completed and to determine whether additional resources need to be deployed.

Today's unions know the importance of properly training and educating their members. This knowledge is career and union based, and the fire service is no different. As new technologies are introduced, firefighters must adapt to the changes, through education and training. The partnership between the human resource department and the labor organization can greatly enhance this process. The labor organization and the human resource management division share a concern over compensation. The fire service administration is responsible for providing the climate, support, and development of employee career planning. However, it falls on the individual firefighter to manage his/her career plan to ensure the greatest rate of return on the investment and ultimately advance his/her career as far as possible.[23]

Recruitment, Selection, and Retention of Personnel

Recruitment is the process by which a department searches for qualified applicants for a position. In larger, full-time fire departments, they may use the media to advertise a position. In volunteer fire companies, they may hold a recruitment drive, by sending current members into the community in search of people willing to serve in the department. The recruitment process establishes an applicant pool from which the department can choose the most qualified candidate. The recruitment process is different from the hiring process. Recruitment, selection, and retention should follow a normal progression to ensure the best possible workforce, as shown in figure 8–6.

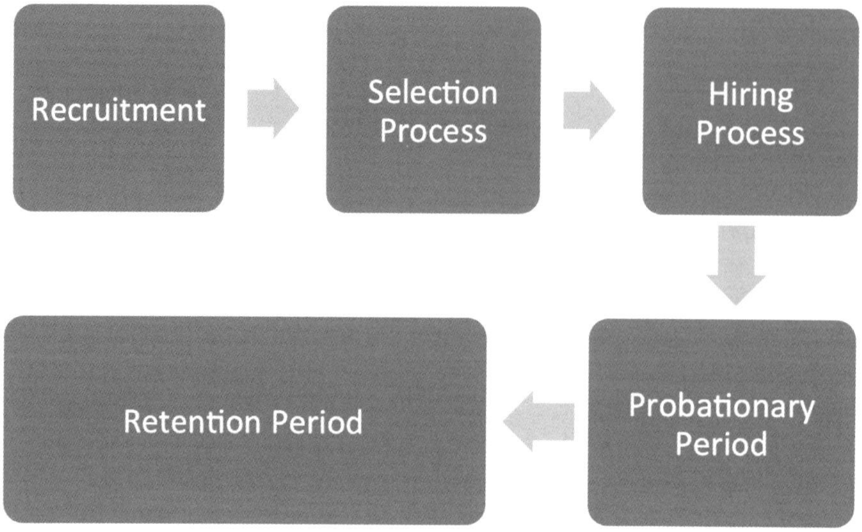

Fig. 8–6. Hiring process

Some volunteer and call fire departments are faced with a very small applicant pool, and consequently, the recruitment process may be an ongoing process for the department. Labor market conditions control the supply and demand for the specific skills for a given job. Full-time fire departments often have a low supply of jobs and a very high supply of applicants, as compared to volunteers departments, which generally have a high supply of positions and a low supply of applicants. It goes without saying that the greater the size of the labor pool is, the better the opportunity the department will have to hire the best applicant.

Some fire departments have a recruitment team that constantly monitors the labor market conditions and determines how the annual recruitment plan should respond. Unfortunately, many fire departments today still use the same recruitment methods they used 10–15 years ago, which, given the economic and social changes in our communities, may no longer be adequate. For example, 20 years ago, a firefighter could join a fire department and would take a 12-hour first-aid class and attend the local fire department training program. Nowadays, firefighters from call, volunteer, and full-time departments across the nation are all required to meet added training requirements, and this has placed a significant strain on recruitment, especially of call and volunteer firefighters. Labor and management should each have the same goal—namely, to hire the best candidate—and to do this, the community must have a pool of qualified candidates. No fire department can function without personnel.

Fire departments may find local job fairs to be beneficial in their search for qualified candidates. Another option is offering an internship within the department to a local college or trade school. An internship is often offered on a short-term basis to a student looking to gain experience in the field as they learn the academic side of the trade. During the internship the student would obtain firsthand knowledge of the job of a firefighter. The fire service can learn from the private sector when it comes to recruitment: what works in one department does not guarantee results in the next department; therefore, recruitment should be viewed as an individual fire department plan.

Some fire departments may be faced with specialized recruitment issues. For example, a department may find that it is not receiving applicants from a diverse population representative of the community. Departments should ensure that the recruitment plan emphasizes diversity.

In 1972, Title VII of the Civil Rights Act of 1964 was amended by the Equal Employment Opportunity Act (EEOA). The EEOA created the Equal Employment Opportunity Commission (EEOC), which has the responsibility for administering the act. Further, state and local governments and public and private educational institutions were included under the act as amended. In addition, the law was strengthened to allow the EEOC to file suit against employers in court for violating the provisions of the act. The EEOC commissioners, any aggrieved person, or anyone acting on behalf of an aggrieved person may file a complaint within 180 days of the last time the alleged act occurred. When a state agency is involved in the complaint, the EEOC filing deadline is extended to 300 days.[24]

The EEOC's goal is to

- Issue employment guidelines and monitor employment practices
- Investigate charges of discrimination and enforce the EEOA

The EEOC's guidelines are not federal laws per se; rather, they are administrative rules and regulations, published in the *Federal Register*. Administrative rules and regulations are enforceable as federal law, even though they are not called laws. Certain federal agencies, such as the EEOC, are authorized to create and pass regulations, and some agencies are authorized to enforce these regulations, which have the same weight as other federal laws. Any administrative agency decision is subject to judicial review from the courts.[25] The courts have given considerable weight to the different rules and regulations while interpreting the law. The EEOA requires organizations to maintain extensive records and detailed reports. In addition, organizations must display equal employment opportunity notices in public view for all employees to see. Many fire departments rely on the human resource management division to ensure compliance with the law.[26]

Title VII of the Civil Rights Act also includes the Americans with Disabilities Act and the Age Discrimination in Employment Act. Both prohibit employers from discriminating based on age, race, national origin, sex, disability, or religion. According to attorney J. Curtis Varone, discrimination is prohibited in[27]

- Testing, hiring, and discipline of employees
- Wages, assignments, or classifications of employees
- Transfers, promotions, layoffs, or recall
- Recruiting and advertisement
- Training and apprenticeships programs

- Benefits, retirement, and disability leave
- Other terms and conditions of employment

It is important that fire departments annually review their advertisement, recruitment, selection, testing, promotional, discipline, and retention programs to ensure that they remain within the scope of the law at the federal and state levels (fig. 8–7). What was once an approved process may no longer be considered legally prudent on the basis of current legal standings.

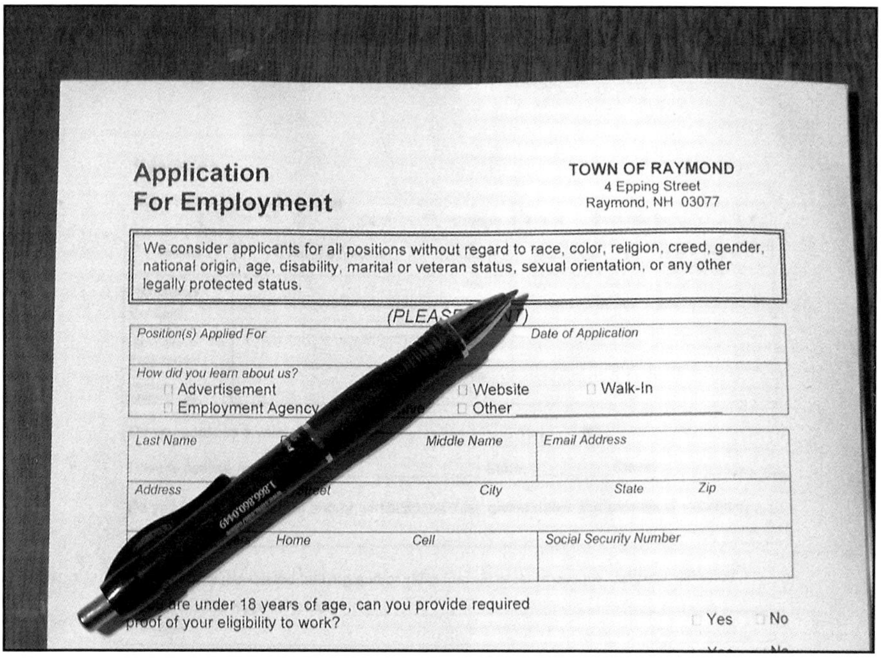

Fig. 8–7. Employment application annual review. Applications should be reviewed on an annual basis to ensure that the application meets all applicable laws.

Once a fire department has established an applicant pool, it can begin the next step, which is the selection process. The selection process involves processing the applications, interviewing qualified candidates, and choosing the best person for the position. No process or program can guarantee that the department hires the best candidate each and every time. Even the best candidate may not be the best employee down the road or may not even fit into the organization. Individuals who apply for positions may be overqualified, underqualified, or have a colorful work history. Selecting the most qualified person now could yield great dividends in months and years down the road for the organization; however, in some cases, departments have selected the only candidate at the time with the hope that they can make the person fit into the position, only to find that this does not happen and the employee eventually leaves—voluntarily or otherwise. This process cannot be rushed, and its importance to the health of the organization should not be underestimated.[28] As Milkovich and

Boudreau have explained, "in most organizations, human resource professionals play a key role in the external selection process. They often recommend and design specific selection techniques, frequently conduct the selection process, and usually help to interpret the results of selection, ensuring that the process meets organizational, employee, and legal requirements."[29]

Many formal selection techniques are used today. Fire departments may utilize a test as one of the means for determining the best candidate. Fire departments may also use psychological screening (Rorschach test, Thematic Apperception Test, and Minnesota Multiphasic Personality Inventory) to determine whether the candidate is psychologically able to perform the job. The psychological evaluation will rate the candidate in terms of the following areas: reaction to stress, flexibility, character, temperament, and personality traits. It is recommended that the fire department combine a number of screening tools in determining the best candidate for the job.

One of the most widely used techniques is the interview process, during which the examiners ask predetermined questions and gather information about candidates from their answers.[30] Fire departments may also require a video interview of the candidate. In this case, the department will send out a set of questions that it would like the candidate to answer in a specific time frame (e.g., 10 minutes), and the candidate can then submit the video to the department electronically or burn it onto a compact disc. This allows the department the flexibility to review the videos at its convenience. This is a low-cost method that can eliminate the need for multiple interviews. For example, the video interview sample is often used in the selection process for fire chiefs, saving the candidate the time and expense of traveling to the community for a first round of interviews.

Once the fire department has completed its selection process and has determined a promising candidate, a formal job offer should be made to the candidate. The job offer is often made by the human resource manager to ensure that all legal requirements are met and to finalize the hiring process from this point forward.

Many local fire departments have a list of applicants who have taken a written exam and received credit for such things as military time served and specialized training. Some states have established civil service lists of eligible candidates. Many communities and civil service requirements require that three or more names of the top candidates be submitted to the requesting official. Generally, a department may be required to hire one of the top three candidates unless it can show valid reasons the candidates are not hirable.

Many union contracts and fire department polices require that a newly hired firefighter must complete a probationary period during which the fire department evaluates the firefighters' ability to perform the job as outlined in the job description. Labor contracts will cover new employees once they are hired by the fire department; however, many collective bargaining agreements contain language that allows the fire department to terminate a probationary firefighter for any reason. Firefighters terminated within the probationary period are often not allowed to use the grievance process under the contract. In some cases, probationary firefighters are not allowed to join the union or are required to pay an agency fee. Departments can choose to extend a probationary period, if they deem the need for further monitoring of the firefighter.[31]

If firefighters on probation do not meet the intended requirements, then they should be dismissed immediately. No cause should be given as to the dismissal, to prevent any legal complications. When employers provide the terminated employee with too much information on why they have been fired, this information can be turned around, landing the employer in a legal battle over the termination. The fire service administrator should seek advice from the human resource manager and the legal department to ensure that the termination is conducted in accordance with the law. Note that you cannot fire a person on probation for anything you want; for example, if you terminate an employee on probation because she is pregnant, then the department would be open to a civil action.

Retention is the process through which a fire department keeps personnel on the job once they have been hired. The retention program has direct links to the recruitment process, because firefighters who chose to join the fire department will choose to stay on the fire department for many of the same reasons they joined in the first place. The retention program often falls to the human resource manager to monitor. Full-time fire departments generally do not have a big issue with retention because most of their firefighters do often not migrate to other departments. Many of the retention programs implemented in fire departments have had limited results, falling short of what was expected from the program (fig. 8–8). If a retention program is going to work, the individual fire department will need to develop its own strategies and tailor the program to the local needs.

Fig. 8–8. Miles City (Mont.) Fire Department retention program. Call and volunteer fire departments need to pay attention to their retention rate.

In full-time fire departments, competitive wages serve as a good retention program, while call and volunteer programs either do not have this option or have very limited resources to offer a competitive wage. Structuring jobs so they are appealing and satisfying is another area that can be used in all fire departments, regardless of their makeup. Developing a strong management team is the third area that all fire departments can use to retain firefighters. Developing opportunities for job growth—such as specialized teams for Hazmat, high-angle rescues, emergency medical services, or maintenance—is the fourth area that can be incorporated into a retention program.[32]

A retention program for your department should take into account the group of employees that you are trying to retain and what drives them. What works for the baby boomers may not work for retaining the millennials. The millennials, born between 1980 and 1995, comprise the next generation of firefighters. On the one hand, they will generally only take yes for an answer and they feel they come first over other things. On the other hand, the millenials are multitaskers who can manipulate any electronic device on the market and thrive on instant gratification. This is polar opposite from the baby boomers.[33]

In between the baby boomers and the millennials are the Gen Xers, who are more than willing to challenge the status quo. This group wants freedom from being micromanaged and is more than willing to balance multiple projects. The Gen Xers may be in conflict with the millennials because of conflicting views and priorities. As outlined, each of these three groups will respond differently to retention efforts. As you read the following chapters, consider how these three groups will respond to the following:

- Bargaining topics/issues
- Progressive discipline
- On-the-job counseling
- Benefits/wage issues
- Dispute resolution

No matter what size or type of fire department, the fire department must conduct research, develop, implement, monitor, and evaluate the retention program. If a fire department currently has a retention program, it is recommended that the program be reevaluated to ensure that the program is producing the intended results.

In line with retention, firefighters today are often responsible for caring for family members. *Family responsibilities discrimination* (FRD) is a fairly new legal challenge facing many public and private sector employers today. Cynthia Thomas Calvert, of the Center for WorkLife Law, has stated that FRD "occurs when an employee suffers an adverse employment action based on unexamined biases about how workers with family care giving responsibilities will or should act, without regard to the workers' actual performance or preferences."[34] Employees are using Title VII of the Civil Rights Act of 1964 to protect family caregivers from claims of disparate treatment, disparate impact, harassment, failure to promote, retaliation, and discrimination based on gender stereotypes. Other sections of the law that may be incorporated into a claim to protect family caregivers are the Family and Medical Leave Act of 1993 and the Pregnancy Discrimination Act of 1975. Fire service administrators must ensure that they stay current on newly issued legal briefs pertaining to FRD. The human resource department should train supervisors in ways to avoid FRD claims when dealing with employees. Supervisors and administrators should be flexible and responsive toward employees' work-life issues.

When employees think or feel that they have been the subject of FRD, it tends to have a negative impact on job commitment and productivity. Dr. Christine Dickson has reported that "chronic work-family conflict has been linked to a variety of medical and psychiatric conditions, such as high blood pressure, high cholesterol, depression, anxiety, and substance abuse, any of which may increase disability and workers' compensation claims.[35] The fire department employee assistance program should providing training and education (for more about work-life balance, see chapter 9). FRD claims now affect all types and sizes of fire departments across the country.

Call and volunteer fire departments can struggle with retention issues. With greater demands being placed on families today, many people work two or more jobs just to survive. Some families comprise single-parent homes. Family obligations place a high demand on some individuals. These are just some of the reasons that people have very limited or no time to volunteer for an organization. This challenge is not unique to the fire service; many volunteer organizations have the same issues. Many private volunteer organizations have aggressive retention programs, which the fire service could use as a model. Today more than ever, the call/volunteer fire department must be flexible, to best meet the needs and wants of its members. There must be open and honest communication within the department, and care should be taken to set a respectful tone in the department, offering praise to its members.

Student Activity

Research and identify what your local fire department has for a *recruitment* and *retention* program. Answer the following questions:

- What are the plan's strengths?
- What are the plan's weaknesses?
- How does your plan meet the legal requirements for recruitment?
- How can this plan be updated (areas for improvement)?
- What is the target population for the recruitment plan?
- What does the department offer for retaining employees?

Job matching is a significant part of the retention aspect for volunteers. Not all volunteers want to work on the ambulance, and not all volunteers want to fight fires. Some people are content volunteering to be a dispatcher, performing general maintenance on the vehicles or the station, preparing the budget, or grant writing. When the department can match the job to the person, the chances of retention are much greater.

One section of employee relations that has gained much attention in the fire service lately is the implementation and utilization of *performance appraisal* systems, which play a major role in the performance management program. Here we will briefly review the concerns and issues facing labor-management relations when dealing with such a program. Edwards has maintained that "performance appraisal is the formal, systematic assessment of how well employees are performing their jobs in relation to established standards and the communication of that assessment to the employee and the organization."[36] Management may view the performance appraisal system as a tool for making sound business decisions and employee performance decisions.[37] Some have argued against the need for performance appraisal systems.[38] The first generation of performance appraisal was a means to determining whether the salary or wage of an individual was justified, whereas today the focus is on providing employees with insight as to how they are doing by use of a predetermined set of criteria. Some fire departments have tied the compensation package to the performance appraisal system. Because the performance appraisal system will have an effect on the firefighter's employment record, labor unions have been either opposed or have strived to be involved in the development of such programs, to protect employees' rights.[39]

People often see the system in a negative light because of past failures, as poorly developed, poorly implemented, or a means for a supervisor to get even with an employee (fig. 8–9). Some evaluators will be reluctant to give a high or low score because they may be challenged to justify the score or may have to write a narrative in support of the grade. The goal should be to eliminate, as much as possible, the ability of the manager to manipulate the evaluation, as well as to remove misguided or even unreasonable evaluations.[40]

The human resource triad is a great example of involving the three key players in the establishment of performance standards to be used in the appraisal process. This creates a balanced approach and allows each section ample opportunity for input. When incorporating all three sections, acceptance by the organization is met with much less resistance. Employees must be given the opportunity to dispute an evaluation without fear of reprisal from the supervisor. The goal of the feedback program should be to have the supervisor and the employee jointly develop suggestions on how best the employee can improve future performance. This dialogue must be ongoing and open to yield the greatest success, opening the door for the supervisor to follow up on the mutually agreed plan of improvement and to see whether the intended goals have been achieved. As you may recall from the "Motivation" section, employees generally want to know how they are doing.

Again, the human resource management division, management, and the labor organization should all be involved in the process. This will ensure that all issues and concerns are being properly addressed to guarantee the best outcome from the system, employee, and the employer. Kavanagh, Benson, and Brown have suggested that the level of employee acceptance to a performance appraisal system comes down to the perceived fairness of the system.[41] Involving the human resource management division and the union in the development will foster transparency, so there is no hidden agenda of either management or labor. Management and labor should also work to ensure justice in the process; employees need to know that they have an appeal process and that the system will be monitored. In his research, Lawrence Nurse found that "there would not be any appreciable differences in justice perception between union members and non-union members."[42] This research debunks the myth that a union-involved performance appraisal process will be given greater review by management.

Fig. 8–9. If properly developed, implemented, and managed, a performance appraisal system can provide employees with feedback on how they are doing and areas requiring improvement.

Final Thoughts

As you have seen in this chapter, effective personnel management practices are no different for the union department or the nonunion department. The goal is to ensure productive resources and quality of service to the community we serve, which we can do only with people and through people. Effective personnel management involves labor relations. Whether you are in control of a small- or a large-scale incident, you are using a management system to manage the incident; as the incident grows, so does your management system. The same is true in your personnel management system; as your human resources grow, your personnel management system should grow to manage the human resources under your command.

The utilization of personnel management is determined by the leadership and the size of the fire department. The fire chief may handle all personnel management issues entirely on his/her own, have a separate community office to handle/advise personnel management issues, and/or seek input from outside resources/consultants. It is best to seek advice from various levels before implementing/acting on personnel issues.

Fire departments that have the benefit of a city or county human resource management office should seek an open and honest line of communication. Some departments may choose to appoint a human resource liaison to act as the conduit between the two agencies. In this case, the fire department human resource liaison may be available to answer technical questions pertaining to the fire department or even act as an advocate for the fire department. No matter how large or small, human resource management issues are an everyday issue for fire departments across the country, and they have a significant impact on labor relations.

Human resource management is not just about getting qualified people hired; rather, it is the entire systems approach, from start to finish. In the process, we must consider what keeps the best firefighters productive and happy. As you have seen, firefighter satisfaction does not always come from the pay or the benefits. The goal of this chapter has been to show the need for human resource management to play a role in the labor relations process. Proper business practices, coupled with human resource management, will build a strong fire department culture that establishes the foundation for firefighters to be productive and happy.

Key Terms

- Activity-authority network
- Age Discrimination in Employment Act
- Americans with Disability Act
- Baby boomers
- Bottom-up forecasting
- Breach of contract
- Controlling
- Culture
- Directing
- Equal Employment Opportunity Act (EEOA)
- Equal Employment Opportunity Commission (EEOC)
- Expectancy theory
- Employee appraisal systems
- Expressed employee contract
- Family responsibilities discrimination (FRD)
- Family and Medical Leave Act
- Federal register
- Gen Xers
- Human resource management
- Human resource management administration
- Human resource triad
- Implied employee contract
- Interaction-orientation role
- Leading
- Management process
- Millennials
- Motivation
- Negligent hiring
- On-boarding
- Organizing
- Performance appraisal system
- Performance management program
- Personality traits
- Personnel management
- Planning
- Pregnancy Discrimination Act
- Recruitment
- Retention
- Selection
- Self-orientation role
- Staffing
- Strategic human resource management
- Task-orientation role
- Zero-based forecasting

Exploring Issues on the Web

HR Professional's Gateway: http://www.hrprosgateway.com/

Human Resource Certification Institute: http://www.hrci.org/

International Association of Human Information Management: http://www.ihrim.org/

International Public Management Association for Human Resources: http://www.ipma-hr.org/

National Human Resources Association: http://www.humanresources.org/

Performance Management Technical Assistance Center, Office of Personnel Management, the Federal Government's Human Resource Agency: http://www.opm.gov/perform/

Professionals in Human Resources Association: http://www.pihra.org/

Society for Human Resource Management: http://www.shrm.org/

Strategic Management of Human Capital, Office of Personnel Management, the Federal Government's Human Resource Agency: http://apps.opm.gov/humancapital/

References

1. Wade, J. W. 2006. Employee retention in the fire service (master's dissertation, California State University, Long Beach). Retrieved September 16, 2008, from *Dissertations & Theses: The Humanities and Social Sciences Collection* database (Publication No. AAT 1431830).
2. Baptiste, N. R. 2008. Tightening the link between employee wellbeing at work and performance: A new dimension for HRM. *Management Decision*. 46 (2): 284–309.
3. Mondy, R. W. 2008. *Human Resource Management*. 10th ed. Upper Saddle River, N.J.: Prentice Hall. P. 105.
4. International City Management Association. 1988. *Managing Fire Services*. 2nd ed. Washington, D.C.: ICMA. P. 77.
5. Dunn, R. 2007. *Haimann's Healthcare Management*. 8th ed. Chicago: Health Administration Press. P. 16.
6. Rohn, H. 2008. Doing more with less: Improve government efficiency and effectiveness through good strategy executed well. *PA Times*. 31 (5): 6–8.
7. Bishnu, S. 2005. Local government organization on its journey to become a learning organization. *Business Process Management Journal*. 11 (4): 388–402. Retrieved June 27, 2008, from *ABI/Inform* database.
8. Congressional Digest. 2004. A different way of organizing government. Congressional Digest. Retrieved June 27, 2008, from *Academic Search Premier* database.
9. *Garcia v. Duffy*, 492 So. 2d 435, 438 (Florida Dist. Ct. App. 1986).
10. Dunn 2007, p. 17.
11. Baribeau, S. 2010. Miami imposes pay and benefit cuts to ease deficit. *Bloomberg Businessweek*. http://www.businessweek.com/news/2010-08-31/miami-imposes-pay-and-benefit-cuts-to-ease-deficit.html (accessed September 10, 2010).
12. Langfield-Smith, K. 1997. Management control systems and strategy: A critical review. *Accounting, Organizations and Society*. 22 (2): 207–32.
13. Starling, G. 1998. *Managing the Public Sector*. 5th ed. Orlando, Fla.: Harcourt Brace and Company. P. 15.
14. Day, D., and R. Lord. 1988. Executive leadership and organizational performance: Suggestions for a new theory and methodology. *Journal of Management*. 14 (3): 453–64.
15. Peters, A., and R. Scott. 2003. What heroes are made of: Firefighter personality—myth or reality. Paper presented at the 7th World Congress on Stress, Trauma and Coping: Baltimore.
16. Antonellis, P. 2006. Burning personalities. *Fire Chief*. 50 (2): 72–76.

17. Niles, B. E. G. 1996. Locus of control, leadership and small group behaviour: An investigation into the socialization of firefighters (master's dissertation, Dalhousie University [Canada]). Retrieved September 16, 2008, from *Dissertations & Theses: The Humanities and Social Sciences Collection* database (Publication No. AAT MM15815).

18. Holborn, R. D. 2002. Motivational factors and personality traits of individuals who decide to enter a career as a firefighter/paramedic (doctoral dissertation, University of Central Florida). Dissertation Abstracts International-B, 63, 2322.

19. Starling 1998, p. 366.

20. Jackson, S., R. Schuler, and S. Werner. 2009. *Managing Human Resources*. 10th ed. Mason, Ohio: South-Western Cengage Learning. P. 322

21. Ibid., p. 315.

22. Ibid., pp. 19–20.

23. Badawy, M. K. 2007. Managing Human Resources. *Research Technology Management*. 50 (4): 56–74.

24. Mondy 2008, p. 72.

25. Varone, J. C. 2007. *Legal Considerations for Fire and Emergency Services*. Clifton Park, N.Y.:Thomson-Delmar Learning.

26. Bohlander, G., and S. Snell. 2010. *Managing Human Resources*. 15th ed. Mason, Ohio: South-Western Cengage Learning. Pp. 124–26.

27. Varone 2007, p. 376

28. Mondy, 2008 p. 162.

29. Milkovich, G. T., and J. W. Boudreau. 1997. *Human Resource Management*. 8th ed. Boston: Irwin McGraw-Hill. P. 238.

30. Bohlander and Snell 2010, pp. 271–77.

31. Fossum, J. A. 2009. *Labor Relations, Development, Structure, Process*. 10th ed. Boston: McGraw-Hill Irwin. P. 125.

32. Fried, B. J., and M. D. Fottler. 2008 *Human Resources in Healthcare: Managing for Success*. 3rd ed. Arlington, Va.: Health Administration Press.

33. Safer, M. 2008. The millennials are coming. *60 Minutes*. May 23. http://www.cbsnews.com/stories/2007/11/08/60minutes/main3475200.shtml (accessed June 10, 2010).

34. Calvert, C. 2010. Families responsibilities discrimination: Litigation update 2010. http://www.worklifelaw.org/pubs/FRDupdate.pdf (accessed July 12, 2010).

35. Dickson, C. 2007. Avoiding family responsibilities discrimination. *Journal of Employee Assistance*. 37 (2): 16–36.

36. Edwards, S. 2005. *Fire Service Personnel Management*. 2nd ed. Upper Saddle River, N.J.: Pearson Prentice Hall.

37. Nurse, L. 2005. Performance appraisal, employee development and organizational justice: Exploring the linkages. *International Journal of Human Resource Management*. 16 (7): 1176–1194.

38. Deming, E. 1986. Out of the Crises. Cambridge, Mass.: MIT Center for Advanced Engineering Study.

39. Carter, H., and E. Rausch. 2008. *Management in the Fire Service*. 4th ed. Sudbury, Mass.: Jones and Bartlett. P. 282.

40. Nurse 2005, p. 1181.

41. Kavanagh, P., J. Benson, and M. Brown. 2007. Understanding performance appraisal fairness. *Asia Pacific Journal of Human Resources*. 45 (2): 132. Retrieved September 17, 2008, from *ABI/Inform* database (Document ID: 1305649731).

42 Nurse 2005, p. 1190.

STRESS AND PSYCHOLOGICAL SUPPORT

9

Learning Objectives

After reading this chapter, you should be able to:

- Describe the importance of having formal employee assistance and wellness programs.
- Distinguish the connection between personnel problems associated with relationship issues and family issues.
- Explain the elements of a wellness program for the fire department.
- Appraise how a fire department responding to a line-of-duty death/ injury will affect the labor relations process.

Today, fire departments, no matter the size or type, cannot escape the fact that firefighters are affected by workplace stress and health concerns. Fire service administrators cannot escape the stress and psychological effects from the job. Fire department members have a high exposure rate to significant incidents, as well as repeated exposure to incidents that most citizens may never experience. So it is not fair to expect the average citizen to understand the extent of emergency service stress and its impact on the organization, the individual firefighter, and the firefighter's family. It is common today to find contract language on employee assistance and other work-life issues. The following chapter will explore issues that can affect the individual employee, such as drug/alcohol use, marital issues, and injury/illness.

The fire department or the labor union is no different than any other organization that develops subcultures built around norms and values, which often shape the attitudes and behaviors of the firefighters who work in the department.[1] Those who marry a firefighter also must adhere to these values and norms; when the spouse of the firefighter is in conflict with the values and norms, marital conflict generally results. All these issues are interconnected with labor relations. In many cases, labor and management can work together in an attempt to minimize the effects of stress on the firefighter and the firefighter's family. The goal of labor and management should be to keep good employees healthy, productive, and on the job.

Marital Issues

The fire service is no different than any other type of business when it comes to domestic abuse and marital issues. According to nationmaster.com, the United States ranks number one for divorce rates per thousand people, some would argue that the divorce rate is even higher for firefighters than for the general populations. As Dr. Portia Rawles, of Regent University in Virginia Beach, has stated, "firefighters have one of the highest divorce rates when compared to other occupations."[2] Most firefighters have a coworker who is divorced or going through a divorce.

Given the higher rate of divorces in the fire service, the fire service should be doing more to deal with the issue. Many leadership or fire officer training programs do not really address the issue of marital problems and the impact they have on the organization; nevertheless, many fire service supervisors are faced with members who are affected by marital issues. The intent of this section is *not* to have a fire department interfere or pry into firefighters' personal lives, but rather to recommend that the fire department establish programs to educate and provide resources for firefighters and their families in need of services.

As a fire chief, I realized that when a new firefighter signed on the dotted line, I was not only hiring the individual firefighter but also taking on the responsibility to the firefighter's family in dealing with the emotional fallout of the fire service. I would submit that most fire chiefs have received a call from a concerned spouse complaining about an issue or voicing concern. In one incident, I recall a firefighter's spouse calling the fire chief several days after a line-of-duty death because of the way the department had handled the emotional toll taken on the firefighter. Spouses and family members should be provided up front with information on what it is like to be a firefighter's family, and this information can be obtained in part from the spouses of those already in the department. In addition, separate informational sheets should be prepared for male spouses and female spouses of firefighters; while certain issues are common to both sexes, the two groups have some significant differences. Some fire departments have used a family night to bring in the new firefighters' families to show them what it is like to be a firefighter, as well as to explain what it will be like to be a firefighter's family. For example, the Washington Township Fire Department in Indianapolis conducts a recruit family night, which Lt. Robert Smith, Ph.D., describes as follows: "Family nights consist of brief presentations and a panel discussion by a group of firefighters' spouses. Recruits' families have an opportunity to ask the spouses specific questions about how their families cope with the fire department lifestyle."[3] This is essential when a family is entering the fire service for the first time and is unfamiliar with the social and cultural standards within the fire department; even though this may not be new information for the third-generation firefighter family, it should still be presented.

A support program can be broken into the following sections:

- The individual firefighter
- The individual firefighter's family

- The fire department
- Fire department family members

Labor and management can work on building a support system for the above four sections. While each section may have its unique issues, some services will overlap. The goal should be to make available services to each of the sections. When such programs are implemented, firefighters feel that the organization cares about them and their family. At the same time, families will see that the organization is concerned about the effect that the job has on them. The program will demonstrate that fire service families have a unique system that cannot be compared to the private sector. The fire department traditions and history will often play a role in determining how the department communicates with families and supports them. Firefighters and families will be more inclined to request services if a program is in place and is supported by both labor and management.

Firefighters have a high degree of commitment and dedication to the fire service. At first, this new lifestyle that the firefighter and family is experiencing may be fun and exciting. Over time, however, the firefighter's spouse may start to feel that the fire department is taking the spouse away, and the family begins to feel abandoned by the firefighter, resulting in resentment of the firefighter and/or the fire department. Some female spouses have referred to the fire department as the firefighter's mistress. For some firefighters, the fire department is a place to escape to and immerse himself/herself in the job and not deal with family issues or the marital disconnect. Therefore, the firefighter must learn to balance the firefighter family with the fire department family. Often the two families are in conflict, leaving the firefighter in the middle, being pulled in opposite directions; the demand for time from each family is equally important in the firefighter's eyes, and if the balance between the two is shifted and unbalanced, then one family will suffer. When one of the family units suffers, it tend to take a toll on the individual firefighter. Over time the resentment can advance into anger from the family that is often suffering. On the one hand, the firefighter who comes in and does his/her job and takes on no other extra responsibilities and never comes back for overtime or incidents may be resented by others within the fire department. On the other hand, the spouse of the firefighter who spends all of his/her time working and training for the fire department may come to resent the department and the firefighter for this level of dedication and commitment.

Firefighters are not excluded from having an abusive spouse or being an abusive spouse. This is not about gender. A strong male firefighter can be abused by a female spouse, and abused firefighters could be in a same-sex marriage. What happens when an abused firefighter spouse calls 9-1-1 for help? Does the abused spouse fear that the responding police officers will side with the firefighter with whom they work and party? Will the spouse use protective programs in a manner to embarrass the firefighter in front of his/her peers? Will the fire department administrators intervene and have charges dropped on the arrested firefighter? These are all situations that a department should be prepared for, with programs to deal with each.

In years past, the influence of working together may have played a role in how the police responded to a domestic violence case involving firefighters. Today, no police officer would put his job on the line to protect a firefighter when it comes to domestic abuse. The fire department must send the right message to the firefighters and the families—that the department takes a zero-tolerance stance on domestic abuse. The awareness program should involve services that are available to the firefighter and/or the family in the event of an abusive relationship or marital problems. The department should prepare brochures; reference Internet sites, referral lists, and books; and conduct educational sessions on identifying an abusive spouse and the services/protection available for victims of abuse. The human resource triad (see chapter 8) would be a great starting point for developing such a program.

Families and firefighters need to be provided with

- Education on spousal abuse, including examples and real-life cases
- Training for first-line officer and fire service administrators on dealing with employees who are abused or abusive
- Instruction on how the system works
- Confirmation that labor and management will support the system/process

When a fire department takes such an approach toward family support, the new family will often feel embraced within the fire department family unit and view the fire department as extended family. The family support program should be designed around the individual department's needs; what has worked in another department is not guaranteed to yield the same results in another. Again, the goal is not to meddle in a firefighter's personal life, but rather to provide the needed referrals to support firefighters and their family. Each family is different in terms of how willing it will be to accept assistance from outside the family.

Station Romance

A station romance can be defined as two fire department employees actively involved in a formal or informal relationship and is often a result of a combination of proximity, familiarity, similar beliefs, and attraction. According to Anderson and Hunsaker, those who work side by side are more prone to develop an interpersonal relationship.[4] Station romances occur in all types of fire departments, and the impact a relationship has on the department depends broadly on the type of romance. On the one hand, when a relationship is based on just sex, is an extramarital affair, or serves as a way for a person to advance his/her career, most fire departments take a negative view, and some departments have strict polices banning such relationships. On the other hand, when two consenting adults enter into a serious or formal relationship, this is viewed differently. Even the formal relationship can affect the organization, so care must be taken to monitor this relationship; the human resource manager should sit down with each person and explain employment rights in detail should the relationship come to an end.

Sociological factors may play a role in fire station romances. Firefighters often work long shifts together (many departments have moved to a 24-hour shift), and overtime (including holidays and special events) and specialized training contribute to their socializing on the job. A fire department may also encourage firefighters to recruit like-minded individuals for the department. The fire department culture, attitudes, and beliefs will often determine how a department responds to, monitors, or tolerates a romance; however, these represent only a sample of factors contributing to the development of romance within an organization.[5]

Much research on workplace romance is based on heterosexual relationships. Nevertheless, today's fire service administrators need to be aware, monitor, and, in some cases, address both heterosexual and homosexual relationships within the department. The other factor may be that, at present, there are more heterosexual than homosexual firefighters.[6]

No matter the size or type of fire department, care should be taken when crafting a policy on fire station romance, and the fire department should seek legal advice on the writing and implementation of the policy. Rather than banning any type of relationships, the department should provide guidelines on managing relationships, as well as training, monitoring, and counseling for employees. Company officers, supervisors, coworkers, and fire service administrators must take a proactive approach on firehouse romances. All fire departments must take a zero-tolerance to any form of retaliation or harassment based on a current or past relationship between to employees.

A number of fire departments have a policy that supervisors cannot date subordinates, and in larger departments, there may be a policy that prohibits training officers from dating recruit firefighters. However, when a department forbids any particular type of interpersonal relationship, members of the department will still engage in such relationships and will simply hide and lie about the relationship. No fire department can control or dictate their employees' emotions. So rather than turning a blind eye to the problems that force people to keep their relationships secret, the department should form an education and awareness program. Ideally, the labor union and management should cooperate to develop a mutually agreed program that provides protection to the individual member, the department, and the union. However, when the two firefighters who have been involved in a relationship each belong to the same union, the union may be faced with representing each person on department charges, and consequently, there may be an issue of fair representation. It is important for labor and management to work together on crafting a program that protects all involved.

Fire service administrators must balance individual employee and workplace rights without intruding on employees' private lives and off-duty activities. Labor unions can find themselves in the middle of a dispute between union members over a firehouse romance that has turned sour and has led to harassment charges; therefore, fire departments must consider the potential of sexual harassment claims.

Fire departments that have chosen not to address this issue have found out the hard way that a past firehouse relationship can split the organization. In some cases, these stories have been played out in the local media, which serves no one and demonstrates that a department does not take this issue seriously.

Student Activity

Prepare a response to the following questions, based on your local fire department.

1. Should fire departments be allowed to prohibit romantic relationships between coworkers in an effort to protect the community from sexual harassment claims?
2. What are the negative results of turning a blind eye to romantic relationships?
3. How could a station romance have positive and/or negative results on the organization? What if any are the moral implications?
4. How would you deal with this issue in your department? Provide reasons supporting your approach.

Suicide

The topic of suicide is one that is seldom discussed in the open and is often misunderstood, but it lurks in the dark shadows of the fire service. The topic of suicide will often vary from culture to culture, department to department, and incident to incident with no clear-cut attitude present. The reality is that suicide does happen in the fire service. Between 2008 and 2009, there were seven suicides within the Chicago Fire Department, while several other Chicago firefighters attempted to take their own lives. Notably, in Chicago, more firefighters die from suicide than as a result of traumatic injury in the line of duty.[7] This is not meant to direct the spotlight on Chicago, but rather to call attention to a topic that affects the entire fire service. According to a news report by *USA Today*, "At least 100 police officers and paramedics committed suicide while on the job between 1992 and 2001, according to an analysis by the Occupational Safety and Health Administration."[8] Many in the fire service have read about reducing the number of firefighters killed in the line of duty. Seldom, if ever, do you read about the increasing high rate of suicides in the fire service. Many attempted suicides and successful suicides are not recorded or reported, and most departments will not rule a suicide as a line-of-duty death. I would submit that suicide is considered the silent killer of fire service personnel and not enough is being done to prevent it or educate and train firefighters on suicide's impact on the fire service.

According to the World Health Organization, "for every person who commits suicide, there are 20 or more who will attempt suicide. The emotional impact for family and friends affected by completed or attempted suicide may last for many years."[9] Since the tragic day of September 11th, several emergency workers who worked in the recovery efforts have committed and many more have attempted suicide.[10] Suicide is a real and present danger for first responders. Labor and management should work together to ensure that the political, social, and work environment is aware of the dangers of suicide for firefighters and encourage mental health assistance for people considered at risk.

There is no single best way to implement a suicide prevention and education program. Some fire departments and personnel may resist talking about suicide because of the very strong opinions, beliefs, or attitudes that people generally hold about the topic. The lack of open communication is part of the reason that suicide is misunderstood, and people are quick to develop a negative view of the topic. The fire service needs to bring forward the issue by implementing an awareness program. Again, it is up to the individual fire department to assess what it needs from the program. Through awareness and education, our focus extends from the people we are called out to help to ourselves and/or our coworkers in need. Incorporating a suicide prevention program could mean the difference between life and death of a firefighter.

Generally, suicidal people demonstrate warning signs that they are at greater risk when they

- End a close relationship
- Anticipate a change in employment (downsizing, forced retirement, demotion, suspension, termination, or possible other employment changes)
- Experience a change in health
- Increase their alcohol and drug abuse
- Are currently depressed
- Have a family history of suicidal behavior/attempts[11]

It is so important that coworkers, family members, supervisors, and friends take any and all clues or warnings seriously and not downplay any warning signs. A very high proportion of people who have committed suicide communicated their intentions in some way to others. Because the idea of suicide is often temporary, immediate intervention is very important and can make the difference between life and death.

Additional Resources

For assistance in preparing a *Suicide Prevention and Crisis Intervention Guidebook*, please visit the San Jose (CA) Fire Department Web site (http://www.firenuggets.com/suicide.htm). You can customize your free 22-page guidebook to meet your individual needs.

Employee Coaching

As a fire service supervisor, you are also in the position of coach to the team of firefighters under your command. An employee may be coached for career development, in certain responsibilities on the job, or in dealing with the physical and psychological impact of the job. Generally, employee coaching involves a formal assessment process to determine the employee's strengths and weaknesses, to recognize those areas needing improvement. The coaching process starts with the fire chief and the officers at the top of the fire department, and municipal agencies sometimes hire professional executives to work with department heads.

Importantly, employee coaching is not about providing knowledge and skills (job training) to perform a given job; rather, it is a continuous process aimed at identifying the employee's strengths, weakness, and challenges so that the employee reaches excellence in his/her career. As the coach, you will provide options, from which the employee may select whichever method is best for them in overcoming a career challenge. Employee coaching is not about having the firefighter sit in your office with you telling them when and how they should do something.

As a supervisor, you may find that one of your employees is performing in an unproductive manner hurtful to himself/herself professional or personally. First, talk with the employee one-on-one in a private setting, because just identifying the issue to the employee can result in a change in behavior. For example, an employee who is being challenged by a financial matter may be taking the issue to work. Having a previous coaching relationship with the employee would allow the supervisor an opportunity to explore different options. Again, it is not about telling employees what to do, but instead providing them with options and empowering them to elect the option suitable to their needs. In the example the financially strained employee, the supervisor could suggest the use of employee assistance and wellness programs, human resource management, or other professional help. In the simplest terms, employee coaching is attempting to identify potential issue/concerns before they become official issues on the job. Early intervention is the best approach. The following sections details areas that may be involved in the process of employee coaching.

Employee Assistance and Wellness Programs

Labor unions have a long history of fighting for employee safety and health issues. Health care costs have continued to increase, with no relief in sight; labor and management should be working together to control the rising costs of health care with the use of *employee assistance programs* (EAPs) and *wellness programs*, which have become commonplace. There are different types of EAPs. EAPs and wellness programs allow employees and their families to choose healthier ways of living, resulting in a decrease in illness and injuries and further resulting in health care savings to the employee and to the employer. Thus, EAPs and wellness programs can benefit both labor and management.

EAPs encompass a range of diverse services. Some of the counseling and advisory services offered include the following:[12]

- Financial advice
- Legal advice
- Child and elder care referral
- Adoption assistance
- Mental health counseling
- Life event planning

Firefighters are faced with physical and emotional challenges throughout their careers. The good news is that a majority of the firefighters will be able to manage these challenges on their own, without outside help. However, firefighters may periodically be faced with challenges (e.g., use/abuse of drugs, alcohol, gambling, financial, stress, and marital problems) for which they may need professional help. A firefighter dealing with such overwhelming challenges can become an unproductive employee for the department, so the goal of the EAP is to keep good people on the job and functioning and with a healthy lifestyle.[13]

The fire department has invested a great deal of resources and money in training and equipping firefighters, so it only makes sense to invest money in keeping the individual firefighters productive and healthy. Unproductive employees can lead to problems for management and labor unions. The benefits of EAPs are extensive in addressing personal and professional challenges.[14]

The fire service has a long tradition of protecting its own. This protective nature can lead us to look the other way when brother or sister firefighters have drug or alcohol problems; we may even enable them, intentionally or unintentionally, as by covering for them. Family, friends, coworkers, and employers should not look the other way in hopes that the trouble will pass; by enabling an individual's destructive behavior, we do a disservice to the person. As a supervisor, when you confront an individual, it is best to be upfront and honest during the meeting, so as not to leave this responsibility to brother or sister firefighters.[15]

EAPs have been offered by many public and private firms across the country. EAPs are not new to the public sector or the fire service, with roots dating back to the early 1900s, when the Northern State Power Company and the R.M. Macy Company offered employees an EAP. EAPs were more widely established in the early 1940s as a result of the impact of alcoholism on white-collar workers.[16] Today many people still view EAPs as only dealing with alcohol-related issues; however, as you will see, EAP services go much further. According to the Bureau of National Affairs, in 1972, the Occupational Programs Office of the Federal Institute on Alcohol Abuse and Alcoholism offered federal grants in an effort to increase EAPs across the country.[17] Ever since, EAPs have flourished in the private sector and the public sector, as shown in figure 9–1.

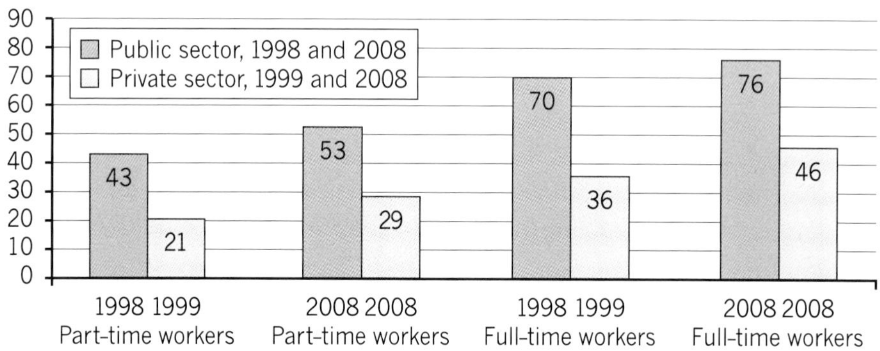

Fig. 9–1. Access to wellness programs and EAPs in the United States. Source: Business of Labor Statistic—Access to Wellness and Employee Assistance Programs in the United States.

EAPs have been tested from time to time, with strongly positive results. Various studies have indicated that, left unchecked, employees' personal and/or work-related problems can have a negative impact on the organization, coworkers, family members, and the individual employee.

The benefits of the EAP are evident on each side of the bargaining table. From the management standpoint, maintaining a healthy workforce will decrease the utilization of health care benefits (e.g., disability and workers' compensation claims), increase motivation, increase morale within the organization, reduce employee turnover, and decrease legal actions from employees; finally, it's just the right thing to do for your employees. From the labor union standpoint, maintaining a healthy membership ensures that the employees are receiving the needed care for personal and/or work-related issues, ensures that members rights are protected, decreases grievance/arbitration filings, and prevents the union from losing members to unaddressed problems. Additional benefits of EAPs include improved job satisfaction, better job performance, enriched work environment, and increased employee loyalty.[18] Today, municipal health and disability insurance programs frequently offer a low-cost or no-cost EAP, in which case, the cost of running the program is embedded in the policy with the municipality; this is done because insurance companies recognize that EAP services reduce claims.[19]

Before implementing an EAP, a municipality must identify which services are sought. The best approach to identifying the services needed by the individual community is to first establish an EAP advisory committee, representing each department within the community, union officials who will be affected by the EAP services, and the employee family members. The following are the recommended responsibilities of the EAP advisory committee:

- Review the needed services and develop EAPs accordingly
- Define implementation procedures for labor and management within the collective barging agreement's terms and conditions
- Identify and evaluate the training/education needed from the EAP provider
- Watch that the municipality health insurance is providing coverage for the services needed by the target population
- Ensure that *all* safeguards are maintained for confidential record keeping
- Provide training and education programs, to employees, supervisors, administrators, and family members, on the use and benefit of the EAP

The EAP advisory committee will *not* have access to who has used the EAP service or any personal information about the program users. The committee will only see generic numbers. Program evaluations conducted by the client after leaving the program will contain only a written form with no identifying comments. This will ensure that the EAP's goals and objectives are being met without comprising confidentiality.

Each section of the committee will bring forward the different needs and wants of a given occupation. The committee should be charged with identifying and listing the required services from an EAP provider. Once the required services have been identified, it will be up to the municipality to determine whether to create an internal EAP or utilize an external EAP. The second question the municipality must consider is whether it can afford all the requirements identified by the committee or whether it should by provide only certain requirements at first, with others to be added as the program grows.

EAPs can provide a wealth of services to a fire department. The level of services being rendered by the EAP will determine the cost of the program. Whereas some communities have expressed interest in only limited EAP services because of the cost, others have been very proactive, requiring that the EAP provider perform in-service training for individual employees, supervisor training, and family awareness programs. In some cases, the EAP is available to communicate with managers about how best to handle an issue in the workplace. EAPs have formed partnerships with labor and management to advance organizational development.

The best practices for an EAP may not necessarily be identical to the needed services for the individual community. Regrettably, many municipalities take a cookie-cutter approach to selecting an EAP provider and, thus, to the services that it offers to workers.

The following are areas for municipalities to consider when exploring the use of an EAP:

- Target problems within the municipality
- Whether the external EAP will accept the municipality's health insurance
- Which services/treatment will be required from the EAP
- Creation of an EAP advisory committee
- Legal issues and collective bargaining issues
- EAP goals, objectives, operations, and marketing
- Program evaluation and feedback methods

Internal EAP

An internal EAP is one that is funded by the municipality, and it is often housed within a municipal building and staffed with municipal employees. In this case, the individual EAP provider is directly employed by the community. Arguably, as an employee of the municipality, the EAP provider will have the municipality's best interest at hand and not necessarily the individual employee's best interest. Additionally, some fear that seeking EAP services from an in-house EAP may result in the leakage of personal information to the municipal leaders of the community or supervisors. A licensed professional EAP provider would not jeopardize his/her position by leaking or giving out protected personal information from a client—not to mention the risk of liability from lawsuits over breach of confidentiality. Still, this remains a fear of many employees who are considering utilizing an EAP and should be addressed up front if the community wants employees to use the EAP. Labor and management should work together to overcome the employees' fears that utilizing an EAP could jeopardize their employment with the municipality.

Strengths and weaknesses of internal EAP[20]

Strengths

- Cost savings for running the program
- Increased utilization rate
- Unlimited visits for employee and family members
- Understanding of municipal organization and dynamics
- Familiarity of municipal policies and procedures
- Better suited for municipalities with more than 500 employees
- Increased input from municipal EAP committee
- Greater control over what services are needed and provided

Weaknesses

- Duplication of services already offered by a private vendor
- Supervisors and senior managers may not use the program for themselves
- Services not offered to meet the diversity of the community
- Ethical conflict (Who is the actual client?)
- Provider is a employee of the municipality

External EAP

External EAPs are based on a contract with a private third-party vendor to provide EAP services to the municipality in return for a fee. In this case, the EAP office may be located within the municipality or a short distance from the city/town. The EAP provider is an employee of the private company and is not an employee of the municipality. Arguably, having a neutral third party running the EAP will ensure a balance representation for the employee seeking assistance and for management.

Strengths and weaknesses of external EAP[21]

Strengths

- Cost of EAP service is often less
- Higher utilization rates for workers, supervisors, and senior managers
- Physical location is not in a municipal building, promoting a feeling of confidentiality
- May have multiple locations and a wider referral list of providers
- Better suited for municipalities with fewer than 500 employees

Weaknesses

- Limited awareness of organizational culture
- Limited awareness of municipal policy guidelines
- Must maintain a profit/loss margin
- Capped services (number of free visits, often three visits)
- Capped supervisory consultations on how best to handle a incident
- Limited oversight from the municipal EAP committee
- Decreased control over what services are needed and provided

Today, EAPs have enhanced the health of the workforce and also have gained credibility with unions and management. EAPs generally offer referral to outside agencies or individuals, or they conduct counseling services within the program for employees. Generally, you will find EAPs that cover topics such as legal, financial, marital, emotional, and drug and alcohol–related problems.[22]

Student Activity

Research which type of EAP you have access to as a first responder for your community. (You may call the EAP to obtain the following information.) Identify the following:

- What model does the EAP use? Internal, external, informal program, or union based?
- What is the physical location of your EAP?
- What does the EAP offer the employee?
- What does the EAP offer the supervisor?
- Does the EAP do counseling, or is it only a referral program?
- How many free visits are you allowed under your EAP? Under your health insurance coverage, how many visits are you allowed?
- In your opinion, what are the strengths and weaknesses of your current EAP?

Release of information. As a point of clarification, the release of any medical or personal information is generally the result of a client signing a release-of-information sheet at the EAP office (fig. 9–2). On several occasions, firefighters have claimed that EAPs released their personal information only to find that the individual firefighters signed a release indicating that the EAP office may provide the specified documents to the municipality. When an individual has signed a release, at any point thereafter they may revoke the release-of-information sheet; in doing so, consultation with a labor attorney is recommended before submitting a written request to the EAP office.

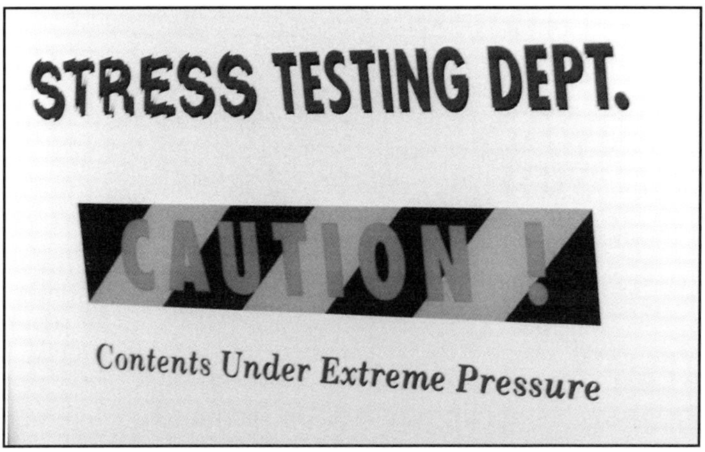

Fig. 9–2. Visiting the EAP office. On the first visit, employees will supply contact and health insurance information and fill out forms including release of medical information. It is important for employees to read and understand all the documents they are asked to sign during the initial meeting at the EAP office.

EAP services may be a condition of employment as the result of a pending disciplinary action. In this case, the release of information may be required in order to fulfill the agreement between the employee and the employer. In such cases, the information released is generally the person's name and the fact that he/she has been attending EAP sessions. In very few cases will a municipality request actual medical records from a particular EAP.

Experts in the field of employee assistance recommend that the EAP provider conduct training for all employees within the municipality on the issue of program confidentiality, record keeping, and the relationship between the EAP provider, client, and the municipality. There is a direct correlation between EAP awareness training and utilization rate: the more active an EAP is in providing awareness training, the greater the utilization rate will be. People need to know that their personal health information will be stored in a secured location with only approved access. Additionally, firefighters and police officers often want to know how long EAP records will be stored and how the records will be destroyed after a case is closed.

Many states have laws that allow the EAP counselor to break confidentiality laws if a life is at stake. For example, Firefighter Bob meets with the EAP counselor at his supervisor direction, but denies any wrongdoing and insists that he is not depressed; he instead claims that his supervisor has it out for him and that is why he is seeing the EAP. Then, two days later, Bob calls the EAP counselor after being out all day drinking and states that he is feeling very sad and wants to commit suicide, identifying how and when he plans to kill himself. Because of the remote location of Bob's home, the EAP counselor calls the local police department to send officers over for a welfare check. In this example, it was appropriate for the EAP counselor to break the very strict confidentiality laws because Bob's life was at stake. During an employee's first visit to the EAP office, the counselor will review such case scenarios with the employee to ensure that the person clearly understands what would trigger a break in confidentiality.

Wellness programs

The foundation of a wellness program is rooted in health education and awareness. Some of the benefits derived from wellness programs are increased worker satisfaction and productivity, decreased stress, decreased disability claims, reduced benefit costs, and improved employee retention rates.[23] By allowing employees to seek professional help from a licensed counselor in addition to referral services to outside providers, EAPs typically address more serious employee concerns than do wellness programs. Nevertheless, a partnership should be formed between the wellness program and the EAP to yield the greatest good. A wellness program is another tool for the organization and the unions to maintain good employee/member health.

A wellness programs works on the principle of preventing illness or injury down the road. The trend toward use of wellness programs is fairly new; still, there are limited longitudinal data to support the program benefits, as well as many recent anecdotal reports attesting that a wellness program can promote a healthier and more productive workforce.[24] Babcock has shown a direct link between stress and chronic illness, with particular increases in costs associated with depression, autoimmune disease, gastrointestinal disorder, and cardiovascular disease.[25] Shannon Klie has put forth that "evidence is mounting linking shift work to cancer, cardiovascular disease, workplace injuries and pregnancy complications."[26]

Labor unions have shown resistance to the implementation of wellness programs. Some of the reluctance that labor unions have to accept wellness programs comes from the following:

- Employee privacy being violated
- Confidentiality of information (shared with insurance company)
- Interference with employees' time off/personal life outside work
- Employer reprisal for employees not taking part in the wellness program
- Time-based fitness programs with punitive results
- Employer concern with bottom-line cost and not wellness of employees

Regardless of their attitudes toward wellness programs, labor or management should both be working to establish a healthier workforce. Generally, poor labor relations between labor and management will yield a poor wellness program; when each side takes the wellness program initiatives halfheartedly, it will not yield the intended positive results. Karla Thorpe, associate director of compensation and industrial relations at the Conference Board of Canada, has stated, "it's really possible for an organization of any size, in any sector, in any part of the country, to invest in a health and wellness program that will have a positive return on investment for an organization."[27] Labor and management need to look at the larger scale of things and make the initiatives for the wellness program that will be in line with the organization's strategic plan and objectives.

A wellness program goes far beyond having fresh fruit delivered to the station once a month. The wellness program should not be a flash-in-the-pan approach. Labor and management both must buy into a wellness program, and it should be supported at various levels within the fire department. The best approach is to survey the members as to what they want and need from a wellness program. It is recommended that a wellness program start off small and build on the needs for the individual fire department. An effective wellness program is crafted by labor and management together on the basis of the individual needs of the fire department's members.

To conduct a quick self-assessment of your fire department, consider the following:

- What physical characteristics does your fire station have to support a wellness program (e.g., vending machines with candy and soda)?
- What physical activities can be conducted in or around your station (e.g., basketball, walking, and stairs)?
- What is the attitude toward fitness among the line firefighters?

- What is their attitude toward nutrition/healthy eating?
- What is the level of physical activity/exercise within the department and when off duty?
- How does management look at fitness?
- What on-the-job injuries could be prevented or reduced with a wellness program?

Note that any personal information complied from an employer wellness program is protected by the Health Insurance Portability and Accountability Act (HIPAA). Before establishing or using a wellness program, inquire as to what type of personal information will be collected, so as to ensure that your wellness program is HIPAA compliant. Involvement of labor and management in this process will help dispel fears about confidentiality issues. It is important to seek input from the firefighters who will be affected by the wellness program and address employees' fears and concerns.

According to the National Compensation Survey, public sector workers have greater access to EAPs and wellness programs than do those in the private sector. Chief among the reasons why the public sector has a higher rate of utilization of EAPs and wellness programs is unionization.[28] According to the Bureau of Labor Statistics, union membership rate for public sector workers stands at approximately 37%, which represents approximately 8% of the total workforce, as shown in figure 9–3.[29] As a result of a high rate of public sector unionization, many collective bargaining agreements currently have articles pertaining to EAPs and wellness programs, as shown in figure 9–4 and 9–5. The labor movement has been instrumental in expanding EAPs and wellness programs.

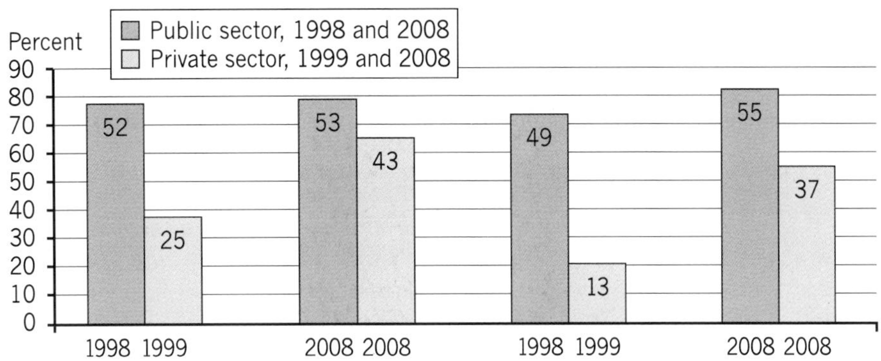

Fig. 9–3. Part-time and full-time workers with access to wellness programs. Source: Business of Labor Statistic—Access to Wellness and Employee Assistance Programs in the United States.

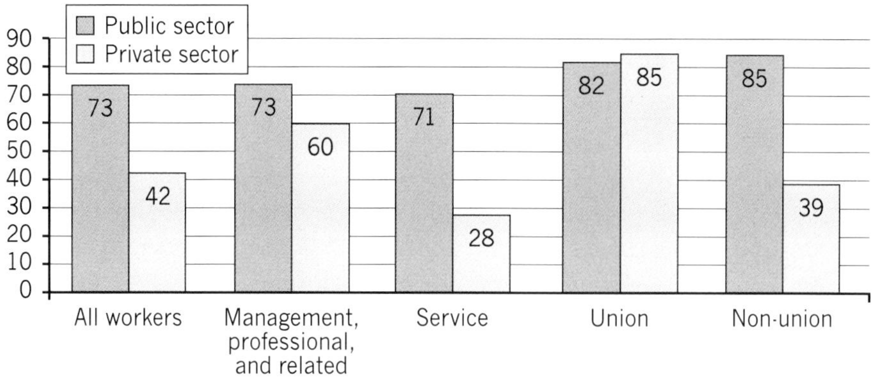

Fig. 9–4. Workers with access to EAPs by selected occupation group and collective bargaining status. Source: Business of Labor Statistic—Access to Wellness and Employee Assistance Programs in the United States.

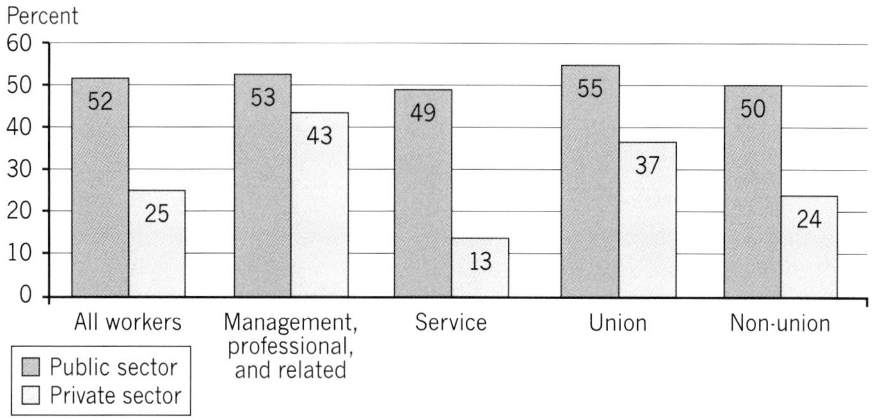

Fig. 9–5. Workers with access to wellness programs by selected occupation group and collective bargaining status. Source: Business of Labor Statistic—Access to Wellness and Employee Assistance Programs in the United States.

Older wellness programs relied on the instructor-led class, handouts, and assignment of a coach to individuals. Today, the shift is toward offering online wellness programs, providing the same handouts along with up-to date information, video samples, and a 24-hour support network to answer questions and provide support. This trend can be of tremendous advantage to the fire service because online programs allow access to multiple stations, 24 hours a day, at a fraction of the cost of having someone physically on location. In addition, firefighters who are off duty may still access the wellness program online. The dissemination of timely information is critical to a wellness program, and online wellness programs allow for immediate notifications. Importantly, the online program is not intended to stand alone; rather, it should be incorporated alongside a traditional wellness program.[30]

Fitness for Duty

The Americans with Disabilities Act (ADA) prohibits employers from requiring a fitness-for-duty exam, unless the employer can demonstrate that the exam is job related. Fitness-for-duty exams can be required because of a physical or emotional injury. When a psychological exam is required, the fire department must shows that the request is based on a concern for the safety of the firefighter, coworkers, or the public, as demonstrated in the case *Coffman v. Indianapolis Fire Department*, No. 08-1646, 7th Cir., 2009. An employee's disabilities and medical history can be kept from the employer if the employee so chooses. The employer must show that the inquiry is job related. Experts in the field have recommended that the questions asked during a fitness-for-duty exam be narrowly tailored to demonstrate "whether the employee is capable of performing the essential functions of his job. Broad questions often run afoul of the law."[31] Generally, the fire department may delay the return of a firefighter who has been on the Family and Medical Leave Act until the firefighter has presented the department with a fitness-for-duty certification provided that the firefighter has been given the required notices.[32]

National Fire Protection Association (NFPA) standards must first be accepted by the fire department as a rule, regulation, procedure, or guidelines to withstand a legal challenge. This includes NFPA 1582, *Standard on Comprehensive Occupational Medical Program for Fire Departments*, which guides a fire department's physician in evaluating a firefighter's ability to perform the essential job tasks. NFPA 1583, *Standard on Health-Related Fitness Programs for Fire Department Members*, should be used as companion to NFPA 1582. Fire service administrators cannot pick and choose which standards they will follow when it is convenient. Each of the standards should be applied objectively. Attorney David Comstock has maintained that "fire department administrators must realize that cases will be resolved by people not familiar with the individual parties who will view a fire department's actions with hindsight in lieu of current legal standards, including those found in the ADA."[33]

A fire department should create a rehabilitation program that supports the following components:

- Agreement between labor and management on job requirements before an injury/illness
- Selection of a medical liaison, understanding job requirements and the fitness-for-duty process
- Selection of a physical therapy business that has specialized training in dealing with firefighter injuries/illnesses
- Establishment of a light-duty program in cooperation with labor and management that will support the injured firefighter during a recovery process
- Establishment of employee reevaluation procedures and specification of frequency of reevaluations

- Acceptance by labor and management of a personalized exercise program for members in the department to identify injury-prone body areas
- Use of a light-duty program to allow firefighters to return to work on a limited basis, according to standards worked out between labor and management
- Creation of an advisory committee to include but not limited to fire department labor/management safety committee members, municipal human resource manager, fire service administration, municipal insurance carrier representative, and municipal safety committee members

After answering the question of whether the fire department can send a firefighter for a fitness-for-duty examination, it must answer the question of what to do next. The simple rule is, if the fitness-for-duty certification says that the employee can perform the work, then the fire department should immediately reinstate the firefighter. The courts have shown concern with regard to how the fire department has handled cases after fitness-for-duty certification has been issued.

The fire department should review the standards and requirements for a fitness-for-duty examination with an attorney specializing in this area (fig. 9–6). Requesting a fitness-for-duty examination is not admitting or conceding that an employee is disabled. The cost for a fitness-for-duty examination is on the requesting fire department.

Fig. 9–6. Some fire departments will break down the topic of fitness for duty into two areas: (1) line-of-duty injury; and (2) employees requesting leave under the Family and Medical Leave Act.

Forced Retirement

As opposed to retirement at the end of a career, forced retirement hinges on who or what terminates the employment relationship and may be the result of an injury, illness, family emergency, or even downsizing. Some reports have estimated that more than 1,500 firefighters and police officers annually receive on-the-job injuries/illnesses that result in their being forced into retirement, with an additional 100,000 firefighters and police officer annually being injured on the job with loss of work-time. Regrettably, many of these people did not plan for a career-ending injury/illness. It is strongly recommended that labor and management work together in educating personnel on the challenges faced by a career-ending injury/illness.

Many human resource management divisions and personnel management systems focus on standard retirement issues but miss the target on forced retirement issues. A normal retirement comes at the end of a long career in the fire service, when the individual firefighter decides on his/her own that the time has come to retire. Dealing with a forced retirement, the firefighter may be looking at retirement years before he/she had planned to retire. The reason behind the forced retirement may not allow the individual firefighter to live his/her retirement as envisioned. Consequently, forced retirement exerts a significant strain on the firefighter and the family.

Because there are so many different disability systems and types of retirement systems, it is impossible to examine every type within the limited space in this book. The following section provides a general overview and observations on the issue of forced retirement, and certain aspects may not be applicable to your department. A forced retirement affects the fire department, the firefighter, the family, and the labor-management relationship. How a forced retirement is handled can demonstrate to the personnel that the department is caring and concerned, or it can show that the department does not care about its personnel. Personnel remaining on the department and their families are all monitoring how the forced retirement is handled, because it may happen to them next.[34]

As a coworker, friend, or family member, understand that the firefighter being forced into retirement has many of the needs identified by Maslow's hierarchy of needs. Maslow's theory is that people would seek the next level of need after acquiring the basic needs. A forced retirement violates many of these needs, further complicating the issue of forced retirement.[35] The hierarchy of needs is as follows:

- *Safety needs.* The firefighter injury/illness violates the need to be safe in the work environment.
- *Social needs.* By no longer having work in common, one can be viewed as damaged goods. A firefighter, once forced into retirement, will have less and less in common with the department. The injury/illness may prevent the firefighter from attending social gatherings. Fellow firefighters may not understand the injury/illness, and the firefighter may feel judged by his/her peers.

- *Esteem needs.* Violation of esteem needs leads to a feeling of being unappreciated and undervalued. Some people become stuck in this level of need, because they develop resentment and bitterness, which can delay the recovery process. The department, coworkers, and family should ensure that the firefighter feels appreciated and valued.
- *Self-actualization needs.* Violation of these leads to a feeling of being unable to reach full career potential.

What was once a positive working relationship between the firefighter and the fire department can quickly turn unproductive as a result of injury/illness. The fire department will want to protect itself and may establish a time line for a return to work status on the basis of the type of injury/illness. For example, the District of Columbia Fire Department policy is that a firefighter has 192 days or 64 work shifts to return to work after an injury/illness. After this time, if the firefighter cannot return to work, he or she must retire and accept a disability pension. By contrast, in the State of Massachusetts, the general rule is to wait a full year to determine whether the employee is returning to work; if not returning, the department will then file an involuntary retirement application on behalf of the employee

There is no national standard to determine the length of time needed to recover from an injury/illness in the fire service. Local fire departments, as well as municipal and state retirement systems, all establish their own time lines for recovery from an occupational injury. This arbitrary practice by which individual fire departments determine the time frame for recovery should be reviewed, and labor, management, medical, and insurance companies should establish a guideline for the recovery process.[36]

When a firefighter is deemed unfit for duty by the fire department fitness-for-duty examination, the case is submitted to the retirement board for consideration. If, however, at the retirement-system medical review, the firefighter is found not to be eligible for disability retirement and can perform the duties of a firefighter, then the firefighter is stuck in the middle; the department examination states he is disabled, yet the retirement-system examination says he is not. One option is for the fire department to terminate the employment relationship at this point, because the firefighter cannot perform the duties. The second option is for the fire department to continue paying the firefighter without allowing him/her back to work. The third option is to allow the firefighter back to full duty, despite the department fitness-for-duty examination results.

Regardless of the option taken, legal action is the preferred response for the firefighter. This type of case is not unusual and more often than not will result in a long legal battle for the firefighter, further compounding the psychological impact of the forced retirement on the firefighter and the family. In general, firefighters who are forced into retirement may experience both psychological and physiological problems, which may show up in the form of depression, unhealthy living, and poor adjustment to the retirement.

When a male firefighter is forced into early retirement and the couple has traditional gender roles, being stuck at home and no longer the breadwinner poses a challenge. Common results include

- Interference with normal routines
- Uneasiness being at home with nothing to do
- Concern about how others will view him/her
- Relationship conflict
- Loss of identity

The retirement process should not be used as an administrative process to remove an unwanted person from the department. Fire administrators need to deal with personnel management issues correctly; forced retirement is not the proper format. Fire service administrators should ensure that the retirement process is used as it was intended, not abused by administration or the individual firefighter. How a department handles a forced retirement can have an impact on the injured firefighter's recovery process and on the organization. If firefighters feel that the department is quick to force people out the door, then they will be reluctant to report line-of-duty injuries/illnesses. For most people, a forced retirement is uncharted water they are sailing on, so provide them with ample information and guidance. The forced retirement is one area of personnel management that is often overlooked because the person is now gone and forgotten by the department. You are strongly encouraged to examine your department and determine how they would handle a forced retirement. What would you recommend to improve the process for your department?

It is not uncommon for people being forced into retirement to suffer depression, alcoholism, drug abuse, gambling, or other addictive behaviors to dull the pain or occupy the free time. In an effort to be proactive, the fire department will need to form a partnership with the retirement system, labor union, municipality, and the insurance company. The fire department should create an environment that is supportive of personnel who have received an occupational injury/illness. The fire service administrators, supervisors, and coworkers should be supportive of members who have been injured. I have seen several examples of the municipal leader and/or the fire chief talking to the news media about an injured firefighter and releasing what type of injury the firefighter has received to the media. In one case, the town manager and the fire chief talked to the media about a firefighter who was out on post-traumatic stress disorder (PTSD) three months after a fellow firefighter died at the fire station when a vehicle fell on him. The release of medical information is a clear violation of the law, and incidents like this create a very negative environment for injured firefighters. A review of the processes of reporting an injury/illness, recovery, and forced retirement should be undertaken by labor and management. As shown in figure 9–7, it is recommended that the fire department follow the forced-retirement dimensions of wellness, a review of the entire process that seeking information from each of the areas on how better to serve the needs of the employee and the department. As addressed previously, the EAP and the wellness program are two options that should be encouraged for the firefighter who is injured/ill.

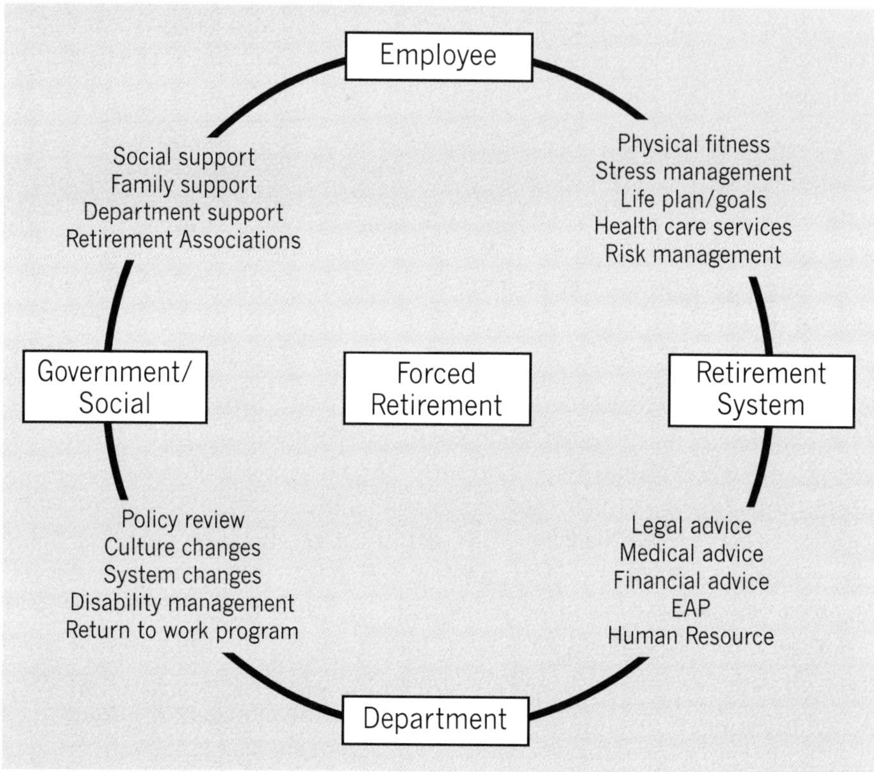

Fig. 9–7. Forced-retirement dimensions of wellness

There is no easy fix for the problem of forced retirement. However, the following steps should be taken by labor and management in moving forward with a forced-retirement issue in the personnel management system:

- Take a proactive approach to ensure that members who have been forced into retirement are cared for emotionally and physically
- Provide awareness and education training
- Involve EAP, critical incident stress management (CISM) teams, human resources, department chaplains
- Develop partnerships and/or a retirement board
- Create an environment that supports a recovery process for injury/illness
- Be respectful and understanding
- Examine data from various locations
- Consider developing/starting a support group
- Encourage continued involvement in the department—employees may be retired but should not be forgotten

Line-of-Duty Death/Injury

There is no question that the vocation of firefighting is extremely dangerous. The U.S. fire service loses on average about 100 firefighters per year and tens of thousands are injured. Between 1990 and 2000, 8% of fatal incidents involved the death of more than one firefighter; these incidents accounted for 18% of firefighter fatalities.[37] This section will examine how a line-of-duty death/injury can play a role in the labor relations process and the personnel management system. Do not underestimate the impact that a line-of-duty death or serious injury can have on the personnel within your department, both in the near term and in the long term. The goal of this section is for you to evaluate your department and how they would handle a line-of-duty death/injury from the moment that it occurs to years later.

How a fire chief handles a line-of-duty death/injury will make a lasting impression on the fire department. The personnel within the department will not recall the good or bad things that have taken place under this administration, only how the death/injury was handled. Fire service administrators and labor organizations should work together on the department policy or procedure for dealing with a line-of-duty death/injury well before it happens.

Many fire departments get lulled into a false sense of security when they have not had a line-of-duty death, imaging that, if they have one, other larger departments will help them (fig. 9–8). While fire departments may have a policy/procedure for dealing with the funeral, a majority of policies/procures overlook the immediate concerns that attend a line-of-duty death. The most common response when this issue is raised is, "The Chief will be on location and can handle it." However, as the fire chief will not always be present when a line-of-duty death occurs, it is important that every company officer right up to the fire chief knows what to do immediately after a line-of-duty death.

Additionally, a number of fire departments do not incorporate into their line-of-duty death policy the days, months, and years after a line-of-duty death; however, the line-of-duty death is now part of the department's history and has long-lasting effects on the organization. How a fire department handles the follow-up after a line-of-duty death funeral will often determine if the change on the organization will be positive or negative. When a fire department does nothing after the funeral has taken place, personnel issues tend to become an issue, even though these may take weeks, months, or even years to develop. Thus, the department cannot afford to return to the day-to-day routine with a false scene of security once the funeral has taken place. The best approach is to be proactive in planning postfuneral activities, as well as providing support to the department and the grieving family.

Firefighters who have witnessed the death or injury of a fellow firefighter and/or assisted in the recovery of a deceased firefighter may experience an acute traumatic stress response. Because of the intense personal loss and trauma associated with a line-of-duty death, the normal grieving process does not apply and can lead the fellow firefighter to experience a delayed grieving process, in addition to developing a traumatic stress disorder. Dr. Hudnall Stamm, of the Veterans Affairs National

Center for Posttraumatic Stress Disorder, has stated that "inattention to the important aspects of normal grief and bereavement in a person struggling with dramatic stress can increase the potential risk for both a traumatic stress disorder in the development of a pathological grief."[38] A firefighter witnesses a line-of-duty death is hit on two fronts: the incident and the bereavement process. Understanding the grieving process that follows a line-of-duty death will better prepare the department for such an incident.

Fig. 9–8. The impact of a line-of-duty death extends far beyond the funeral.

The Grief Recovery Institute has reported that U.S. companies will lose more than $75 billion annually to "hidden grief." It also found 90% of those in physical jobs indicated a much higher incidence of physical injuries caused by reduced concentration.[39] Many private and public sector organizations provide bereavement leave for employees who have experienced a loss. From a social standpoint, organizations provide this benefit to allow people to be with family and friends during the death of a loved one. Based on research, we also know that people who have experienced a loss and return to work too soon have a much higher injury rate. Bereavement leave often ranges from one to five days off, depending on the employee's relationship to the deceased. People who have lost a significant other can tell you that there is not magic number of days off for every person. Rather, some individuals may want to return to work as soon as possible to resume a normal process after the death. Others may need additional time away from the job to process and deal with the significant loss.

Along the same lines, a fire department should be consider how many days off to provide employees who have experienced the traumatic death of a coworker in the fire department. Because very few fire departments choose to address this need beforehand, they are left attempting to answer it immediately after a line-of-duty death; however, this is not the proper time because such decisions are often based on emotions, rather than a solid decision-making process. Many fire departments assert that they will give firefighters time off, if needed, raising the question of what and who will decide if a firefighter needs time off after a line-of-duty death. How will the time off be deducted from the firefighter's time? Will the firefighter be required to use sick time or vacation time, or will the department provide leave without any loss of paid time and service for the employee?

By working together before an incident, labor and management can craft guidelines to determine time off for members of the department affected by a line-of-duty death. Establishing guidelines sends the message to personnel that the department is concerned with the well-being of its members and realizes that a line-of-duty death will have an emotional impact on coworkers within the department. The guidelines will also allow labor and management to see that the benefit of time off is available; when a department has known benefits available, an employee experiencing traumatic grief is more likely to recognize the need for and to ask for time off. Besides, the firefighter who reports to work in such a state of mind will not give the job 100% of his/her attention, which can result in an injury to the individual firefighter or to his/her coworkers; firefighters returning to work immediately after experiencing such a traumatic stress place everyone as risk. Therefore, employees need to feel that they are supported by the department so that they will not seek their own solutions when they are having difficulty making decisions because of traumatic stress, leading to actions that may not be in the best interest of the firefighter or the department.

The fire service prides itself on showing a united front during a line-of-duty death. Firefighters will travel from across the country to show their respects to the firefighter and the family. The fire service does a wonderful job when it comes to a line-of-duty death funeral; however, what we tend to miss is who the organization needs to respond to after the funeral. Many people feel that the funeral will bring closure to the incident, only to find that the emotional wound travels much deeper and does not go away after the funeral. How a fire department responds to a line-of-duty death/injury will have a profound impact on the labor relations process.

Given the high cost of lost productivity associated with grief and the higher rate of injury associated with people returning to work soon after a death, it would be well served if the fire service examined this issue.

Student Activity

Draft a written memo to the fire chief outlining the pros and cons of establishing a formal guideline for dealing with personnel of their department who have experienced a line-of-duty death. The goal of the memo is to properly evaluate the risk and benefits for offering bereavement leave guidelines immediately after a line-of-duty death for firefighters. You should answer the following:

1. What are the benefits of establishing department bereavement leave guidelines for a line-of-duty death?
2. What are the risks associated with establishing such a policy in your department?
3. What resistance would such a guideline receive within your organization and from outside stakeholders?
4. What support would the guideline receive within your organization and from outside stakeholders?
5. In your own words, what recommendation would you suggest for these guidelines and why?

Dealing with the emotional impact within the department is not the only concern for a fire service administrator. According to *A Guide to Help the Fire Service Prepare for a Line of Duty Death/Injury*, "It is not sufficient to support only the person in uniform. Instead, it is necessary to support the entire environment in which the uniformed person operates."[40] A line-of-duty death/injury can have a significant impact on the families of the members of the department. How the department deals with the families will have a direct impact on the entire department. Those fire departments that have included the firefighter and his/her family in the process have seen positive results. The grieving process is moved from an exclusive, firefighter-only process to an all-inclusive, firefighter-and-family process. Information on grief and grief support is disseminated to everyone. People are allowed to talk openly about how this death has affected them professionally and personally. Including the family in the funeral process and the emotional recovery process will yield a better fire department.

Fire administrators must be concerned with how the line-of-duty death/injury will affect the organization as a whole, the individual firefighter, and the firefighter's family. These areas each play a significant role in how the organization will move forward following a death/injury. As you can see, a line-of-duty death/injury can affect the personnel management system. Organizations left to respond on their own will often experience problems in other areas. An organization may find a greater utilization of sick-time usage, decreased morale, and increase in unacceptable behavior by personnel over time.

Final Thoughts

This chapter has explored how firefighters' personal relationship issues, including station romances, can play a role in personnel management in the fire service. These issues receive relatively little attention in first-line supervisor training programs, yet many first-line supervisors will spend a great deal of their time addressing these issues, without the benefit of proper training, education, and awareness. Turning a blind eye to the problem is no longer an option; dealing with the problem immediately and professionally is the best course of action.

Personnel management involves dealing with humans, as well as the emotional and physical effects on humans under different stresses. Fire service administrators and labor leaders need to be understanding and supportive of the various programs available to firefighters. As you have seen, the stress of the job can impact the individual firefighter and, at the same time, provide secondary trauma/stress on the family unit. When the firefighter and or the family does not see a decrease in the level of stress, seeking professional assistance is the best course of action. Importantly, many firefighters will be able to process the stress on their own and see a decrease of symptoms; those who do not see a decrease in symptoms should not feel that they are any less of a person.

Occupational stress left unattended can manifest into larger problems. Firefighters have a great deal of pride, integrity, and respect, so asking for help is almost always a challenge. Often, they may feel that asking for help is a sign of weakness. Suicide has been identified as an area of concern for firefighters, law enforcement, and emergency medical service providers. Additional research is need to track and report incidents of suicide among all first responders. The fire service needs to incorporate suicide training, education, and awareness for all personnel. Additionally, the training, education, and awareness program must incorporate the families of firefighters.

Labor and management must be supportive and anticipate not only these needs but also the reluctance to ask for help. An environment that is supportive of its employees will yield a much greater level of commitment and dedication to the department. Departments who demonstrate a negative opinion of human challenges such as alcohol, drug abuse/use, gambling, fiscal, and emotional and/or physical injury/illness tend to have slower return-to-work rates, increased turnover, and forced retirements.

It goes without saying that part of personnel management is dealing with human issues. Many fire departments across the country have seen positive results from EAPs and wellness programs. There is no single method that is best for every organization; the organization needs to determine what program will be the best for them based on the needs of their employees. In addition, how a department processes a forced retirement will affect the labor/management relationship; involving the different key stakeholders in the process will improve the relationship between labor and management. Providing a customized EAP and wellness program will provide the best possible results for the department and the firefighters, and maintaining a healthy workforce should be a shared goal of labor and management.

A line-of-duty death/injury can also affect the labor relations process. There is no excuse for any fire department in the country not to have a detailed policy or guidelines for dealing with a line-of-duty death/injury! This will increase the odds that the death/injury is handled in the best possible manner and that the emotional and physical needs of the organization, firefighters, and families are met. Coordination of the many events following the line-of-duty death/injury of a firefighter is an extremely important yet complex responsibility. Those who have studied the best practice following a line-of-duty death/injury recommend that professionalism and compassion be exhibited at all times, as well as an obligation to the firefighter's survivors and to the fire service community.

Key Terms

- Americans with Disability Act
- Confidentiality of information
- Employee Assistance Program (EAP)
- Employee coaching
- Equal Employment Opportunity Commission (EEOC)
- External EAP
- Fitness for duty
- Forced retirement
- Forced retirement dimensions of wellness
- Health Insurance Portability and Accountability Act (HIPPA)
- Internal EAP
- Light-duty program
- Line-of-duty death (LODD)
- Release of information
- Standard on Comprehensive Occupational Medical Program for Fire Departments (NFPA1582)
- Standard on Health Related Fitness Programs for Fire Departments (NFPA 1583)
- Sociological factors
- Suicide
- Support programs
- Wellness program
- Work-life issues

Exploring Issues on the Web

The American Academy of Experts in Traumatic Stress: http://www.aaets.org/index.html

The Firefighter's Guide to Managing Stress: http://www.management-advantage.com/products/firestress-book.htm

Fire Service Joint Labor Management Wellness-Fitness Initiative Resource: http://www.iaff.org/hs/wfiresource/default.html

IAFF Online Behavioral Wellness Manual: http://www.iaff.org/ET/JobAid/EAP/eap.htm

International Critical Incident Stress Foundation: http://www.icisf.org/

National Institute of Mental Health: http://www.nimh.nih.gov/index.shtml

National Center for Posttraumatic Stress Disorder: http://www.ptsd.va.gov/

North American Fire Fighter Veterans Network: http://www.firefighterveteran.com/

Phoenix Fire Department: Alcohol and Substance Abuse Education, Awareness, Treatment, and Prevention Program: http://www.iaff.org/hs/wfiresource/PhoenixDocs/Behavioral%20 Health/AlcoholDrugEAPPhoenix.pdf

Preventing Suicide a Resource for Police, Firefighters and Other First Line Responders: http://whqlibdoc.who.int/publications/2009/9789241598439_eng.pdf

Recognizing EMTs, Firefighters, or Police Officers with Post-traumatic Stress Disorder: http://www.ptsdsupport.net/emt.html

References

1. Cooper, C. L. 1998. *Theories of Organizational Stress*. Manchester, U.K.: Oxford University Press.
2. Rawles, P. 2003. Don't divorce yourself from members' marital woes. http://firechief.com/leadership/management-administration/firefighting_dont_divorce_yourself/ (accessed June 16, 2010).
3. Smith, R. 2002. How to balance the two families of firefighting. http://firechief.com/mag/firefighting_balance_two_families/ (accessed June 16, 2010) .
4. Anderson, C. I., and P. L. Hunsaker. 1985. Why there's romance at the office and why it's everyone's problem. *Personnel*. 67 (2): 57–63.
5. Gautier, C. 2007. Managing romance in the workplace. *Journal of Employee Assistance*. 37 (1): 7–9.
6. McIntyre, K. S. 2000. The office: The place to look for love. *Vantage 2000*. http://www.siop.org/tip/backissues/tipjan98/hartel.aspx (accessed June 16, 2010).
7. Blakley, D. 2009. Suicide among police, firefighters is on the rise. *Chicago News*. http://cbs2chicago.com/local/Suicide.Police.Firefighters.2.1008190.html (accessed July 30, 2010).

9. Hopkins, Jim and Charisse Jones (2003). Disturbing legacy of rescue: suicide. *USA Today*. Retrieved November 20, 2011, from http://www.usatoday.com/news/nation/2003-09-22-legacy-usat_x.htm.

9. World Health Organization. 2009. Preventing suicide: A resource for police, firefighters, and other first line responders. http://whqlibdoc.who.int/publications/2009/9789241598439_eng.pdf (accessed July 30, 2010). P. 7.

10. Blakley 2009.
11. World Health Organization 2009.
12. Dessler, G. 2008. *Human Resource Management*. 11th ed. Upper Saddle River, N.J.: Pearson.
13. Stack, D., and P. Antonellis. 2006. Help me, help you. *FireRescue*. November: 78–82.
14. Smith, R. 2002. Big brothers, big sisters of the fire service. Fire Chief 46(6): 27.
15. Stack and Antonellis 2006.
16. New York State Work-Life Services. 2006. History of EAP. http://www.worklife.state.ny.us/eap/history.html (accessed September 11, 2009).

17. Bureau of National Affairs. 1987. *Employee Assistance Programs: Benefits, Problems, and Prospects*. Washington, D.C.: BNA Special Report.

18. Valentine, S. 2004. Employment counseling and organizational ethical values. *Journal of Employment Counseling*. 4: 146–55.

19. Burke, J. 2008. The lessons of "free" EAPs: free and low-cost EAPs are symptoms of predictable business changes taking place within the employee assistance profession. *Journal of Employee Assistance*. 38 (2): 17–18.

20. Csiernik, R., D. Hannah, and J. Pendor. 2006. Review of the University of Saskatchewan EAP. http://www.usask.ca/eap/reviewreport.pdf (accessed July 12, 2010).

21. Ibid.

22. Stoltzfus, E. 2009. Access to wellness and employee assistance programs in the United States. Bureau of Labor Services. http://www.bls.gov/opub/cwc/cm20090416ar01p1.htm#glossary (September 11, 2009).

23. Bossi, M. 2008. Canadian firms shouldn't follow U.S. lead. *Canadian HR Reporter*. 21(8): 29.

24. Fried, B., and M. D. Fottler. 2008. *Human Resources in Healthcare: Managing for Success*. 3rd ed. Arlington, Va.: Health Administration Press. P. 333.

25. Babcock, P. 2009. Workplace stress? Deal with it! *HR Magazine*. 54 (5) 67–72.

26. Klie, S. 2010. Shift work presents many health dangers: Symposium. *Canadian HR Reporter*. 23 (10): 1.

27. Klie, S. 2010. Wellness sees positive returns: Studies. *Canadian HR Reporter*. 23 (8): 8, 12.

28. U.S. Department of Labor, Bureau of Labor Statistics. 2009. National Compensation Survey. http://www.bls.gov/ncs/ (accessed September 11, 2009).

29. U.S. Department of Labor, Bureau of Labor Statistics. 2008. Union members summary. http://www.bls.gov/news.release/union2.nr0.htm (accessed September 11, 2009).

30. Arnold, J. 2009. Putting wellness online. *HR Magazine*. 54 (8): 63–65.

31. HR Specialist. 2010. When determining fitness for duty, strictly limit medical inquiries to essential functions. *California Employment Law*. 4 (7): 2.

32. 29 U.S.C. § 2614(a)(4); 29 C.F.R. § 825.310(a).

33. Comstock, D. 2008. Beware of side effects from fit-for-duty tests. http://firechief.com/leadership/management-administration/beware_side_effects_0908/ (accessed July 13, 2010).

34. Antonellis, P. 2007. Coping with the Challenges of Forced Retirement. *Fire Engineering*. August. Pp. 91–92, 94, 96, 98–100.

35. NetMBA. n.d. Maslow's hierarchy of needs. http://www.netmba.com/mgmt/ob/motivation/maslow/ (accessed July 28, 2010).

36. Thompson, J. 2008. Forced retirement policies in spotlight after D.C. injuries. *FireRescue News*. http://www.firerescue1.com/health/articles/334211-Forced-retirement-policies-in-spotlight-after-D-C-injuries (accessed July 28, 2010).

37. U.S. Fire Administration. n.d. Historical overview. http://www.usfa.dhs.gov/fireservice/fatalities/statistics/history.shtm (accessed July 28, 2010).

38. Figley, Charles. 1999. *Traumatology of Grieving: Conceptual, Theoretical, and Treatment Foundation*. Ann Arbor, Mich.: Edwards Brother. P. 16.

39. Grief Recovery Institute. 2003. *The Hidden Annual Costs of Grief in America's Workplace 2003*. Sherman Oaks, Calif.: self-published.

40. Antonellis, P. 2000. *A Guide to Help the Fire Service Prepare for a Line of Duty Death/Injury*. 2nd ed. Salisbury, Mass.: self-published. P. 14.

TOMORROW'S LABOR RELATIONS 10

Learning Objectives

After reading this chapter, you should be able to:

- Explain the importance of planning for future developments in labor relations.
- Describe how power, membership, and ownership will play a role in the future of labor relations.
- Identify factors that labor and management should consider when planning for the future in the fire service and personnel management.
- Recognize how ethics and fire service culture will play a role in future labor relations.

Today's Labor Relations

As you have read in the earlier chapters, labor unions and the labor movement have played a major part in the history of the United States, advocating for labor laws that provided protection to all Americans. Because many of the advances from the labor movement benefit union and nonunion workers, all Americans enjoy better working conditions today. Through labor laws, labor unions have afforded future protection to workers, and these laws have helped sharpen today's personnel management systems. Equally important, these laws that provide protection to workers are constantly under attack, in an effort to do away with the law or reduce the significance of the law; therefore, from a historical standpoint, organized labor not only has helped pass the laws that many of Americans enjoy today but has also provided protection from legislative amendments to adjust these laws.

Without organized labor, who would take the lead on to provide legislative protection for the current labor laws on the books? Without advocacy, labor laws at the state and national levels would eventually be rewritten—in most cases, in favor of the employer. The labor movement and management will need to work together to reclaim the competitive edge once held by the United States and create a workforce that is competitive in today's labor market. This may slow the tide of American jobs now being outsourced to other countries.

Today, in spite of better labor laws, the labor movement (private and public sector) is facing multiple challenges on a national front, greater emphasis on company human resources, stronger personnel management systems, decrease in membership, and market demands (globalization). Proponents of the labor movement would be quick to admit that the labor movement has done little to develop or maintain a vibrant labor movement. The labor movement has struggled—and will continue to struggle—with balancing:

- Efficiency
- Equity
- Voice

This is evident within some trades unions and plays out at a national and international level.

The following sections will explore what labor relations may look like in the future from a variety of standpoints: national labor union, fire service, individual union member, and aspiring fire service administrator or union officer. As you have noticed throughout this book, labor relations makes up a section of the human resource management field of study; personnel management plays an equal role in labor relations and human resources. This book has focus on the intersections of human resource management, labor relations and personnel management, while recognizing that they are individual yet interrelated areas.

Aspiring fire service leaders and labor union officers will all play a role in the labor relation process and personnel management. No matter how small or how large your role is in the fire service or personnel administration, your actions will have an impact on the labor relations process. Much of the leadership and management skills developed today will have an impact on how the fire service and the labor relations process advances in the coming years.

Power, membership, ownership struggle

On a national level, the labor movement has struggled with the ideas that it should be focusing on—additional /maintaining networking (building social media formats, creating allegiance with other groups/organizations), building a stronger union voice (power), building/maintaining the social aspects of the labor movement (membership), and empowering the members to take a more active role (member ownership). When the labor union can open up the lines of communication to all levels within the organization and allow input from various levels, promising results will be obtained. However, there are different ideas as to what the focus should be for the labor movement: Should the labor movement fight to maintain what it currently has, or should it focus on expanding the membership? On a national level, will trade unions continue to work in a partnership during contract negotiations to support a unified standard for conditions of employment? Or will trade unions move to a more centralized process and return to bargaining independently?

Returning to having one collective voice speaking on behalf of labor may be a good idea. As the old saying goes, there is strength in numbers. However, the one-size-fits-all approach to the labor movement may not be the right direction tomorrow. The same could be said for the fire service, if the labor union and the call/volunteer organizations could all come to the table with one collective voice, the fire service may be one of the most powerful organizations in America. Regrettably, each group has different priorities and plans, with differences rooted in egos, traditions, politics, or, simply put, a differing of opinions. It is my hope that in the future more can be accomplished toward forming a neutral and productive movement for the fire service in cooperation with all organizations, regardless whether made up of call, volunteer, or career firefighters. If you are an up-and-coming fire service administrator, please keep in mind this idea of a unified voice for the fire service as you advance in your career.

As you can see, there is no easy answer—just a lot of differing opinions on the topic of power, membership, and ownership. The employment relationship will change with the ebb and flow of the job market, as well as with social and economic changes, and the goal of the labor movement is to stay ahead of these changes. Responding to these changes is often an indication that we are behind the times; hence, we are a reactive organization, devoting more time and resources to responding to changes, as opposed to a proactive organization, anticipating impending changes. It will be interesting the see what the labor movement will look like 10 or even 25 years from now.

What guided the labor movement 25 years ago can no longer serve as a beacon for the labor movement today. The workforce demographics have changed, and the labor movement needs to adjust to changing demographics of ethnicity, gender, and education.[1] The profile of a union member 20 years ago is not typical of today's union member. As James Parks has stated, "the future of the union movement depends upon our ability to recruit and promote people of color and women, the fastest growing groups of union members."[2] Today's unions must fight the stereotype of union members as bullies or impulsive troglodytes. Changing the common people's view of the old labor unions will not be easy. The millennial generation have little basis on which to build an opinion of the labor movement, other than the stories of family members. On a local and national front, the labor movement must continue to reshape the image of the labor unions of the past.

Today, younger people have been told repeatedly that they—and only they—control their destiny. These people live by wheeling and dealing in the pursuit of the biggest gains. They have been told they need to be entrepreneurs to survive. It is this group, the millenial generation, that the labor movement must adjust to or else miss out on recruitment of new union members from this generation. The goals of the fire service are best achieved when consideration is given to the goals of the firefighters currently serving in the fire service; true leadership take places only when the goals of the firefighters and of the fire service (organization) can thus be aligned. However, in the real world, seldom will the goals of the fire service, labor organizations, and the individual firefighters be the same. For example, ask people 18–25 years old what they know about labor unions, whether they would consider joining one, or whether labor unions are needed today. There are social preconceptions that the labor movement must deal with via education and awareness, realizing that this generation wants a say (ownership) in how things are going to take place that affect them. Members of the millennial generation will not tolerate a labor union telling them how they should vote or how things will be done on the job; the millennial generation has been brought up to question authority and not accept the status quo. As an aspiring fire service leader, it is your job to educate yourself on what motivates the new generation of the fire service, because what motivated the baby boomers will no longer work. Because many of us in the fire service have learned our leadership skills from previous leaders in the fire service, we have to be aware of what this new generation wants and needs, to develop a leadership style conducive to the employees you will be leading (fig. 10–1).

Fig. 10–1. Leadership training. As an aspiring fire service leader, it is your job to educate yourself on what motivates the new generation of the fire service. Attending national and regional training and education programs, such as the Fire Department Instructors Conference, is important.

To understand leadership, you must examine its premises, as well as cultural beliefs and the existing knowledge base. Only then can you move forward to shape your leadership style. In my career, I invariably attempted to prepare myself for the next promotion, thinking at the time that I was ready; once promoted, though, I found out that my real learning for the position was about to begin. I found this to be a theme as I advanced in the fire service and my academic career. So, no matter how much you prepare for advancement, you still have much to learn. If you advance and you fail to learn from your new position, then you set the stage for failure. As the saying goes, moving up the career ladder does not confer competence—the position one holds does not equal ability, and this applies to the fire service and to the labor union. Collectively, we must work to put the right people in the right positions to advance in a positive way.

Looking forward. The labor movement has passed the test of time and the test of falling membership. Not many businesses can claim to have stayed operational as long as many labor unions have. I salute the labor unions that have carried on during the recent downturn, while many businesses have come and gone.

The labor movement remains an active player in the political aspects of labor relations. On a national front, the labor movement must consider which nontraditional industries to target in the future. Still, how professionals will feel about union representation is uncertain; hence, the labor movement should be considering various avenues to building a stronger, bolder, vibrant union today. The following pages will discuss the future of the labor movement from a national standpoint, a local standpoint, and that of the individual union member—none of which can exist without the support of the others.

At the local level, public sector union membership can be directly tied to the passage of laws that allow and support collective bargaining in the public sector as outlined in the early part of this book. Arguably, the labor movement has a direct impact on the economy, to some degree at least. If you will recall from the first two chapters, the public sector of the labor movement has witnessed continual growth in membership. Organized labor has a strong hold on the public sector. The National Education Association, the American Federation of State, County and Municipal Employees, and the American Federation of Teachers are some of the larger public sector unions, together representing 7.9 million public sector employees across the United States. This compares to the 7.4 million union members in the private sectors.[3] Given that local government workers comprise the largest contingent of the public sector membership, we may safely assume that organized labor will be part of municipal and state government for some time. It is important that public management programs incorporate training and education in the area of labor relations.

Local government must set the standard when it comes to training and educating supervisors how best to work with labor unions. This alone can advance the labor relationship considerably, as all too often supervisors are not provided with ample training and education as to how best to work with organized labor and supervisors may even have preconceived notions about organized labor. While the negative stereotypes may be true in some union environments, there are many more union shops that do not hold up to this negative reflection of the labor movement; therefore, such preconceived notions may be detrimental to the manager in dealing with the labor union. Over the past few years, I have been hired as a consultant to provide officer development training to a number of fire departments to help aspiring, new, and senior officers to better understand the dynamics of leadership in today's fire service. Oftentimes the training leads to a stronger relationship between the officer and the administration, because the lines of communication have been opened and the boundaries of leadership have been clearly identified, allowing each side to grow as they move forward.

One can surmise that the trend of growth in union membership in the public sector will continue. As a result, labor relation classes should become a requirement for future public administrators. Many public administration higher-education degrees require management and leadership course work. As a result of the high unionization rate in the public sector, it is recommended that labor relations course work be incorporated into degree programs; in fact labor relations should be a standalone core course in any fire science degree program.

Being a card-carrying union member for many years does not mean that you understand the labor relations process. In the more than 30 labor relations courses that I have taught over the past five years at the graduate and undergraduate levels, the majority of people who have taken labor relations and belonged to a labor union walked away with a much greater appreciation and understanding of the labor relations process. Fire service leadership training programs should be incorporating sections on labor relations for senior fire officer training programs. As stated above, the position one holds does not equal ability.

Today, the fire service should be training all mid- to senior-level fire service administrators to deal with labor relation issues (or for that matter human resource issues). Dealing with labor relations and human resource management is no longer considered a fire service administrator's responsibility; rather, the duties fall on the first-line supervisors. We need to train supervisors; give them authority, accountability, and responsibility; clarify their roles and performance expectations; and state the measurements and the milestones. All too often, supervisors point the finger up the chain of command when in fact the person best able to deal effectively with the issue at hand is the first-line supervisor.

Today's labor union is often viewed in a sphere of limitations, and it needs to focus on developing a value that it can add to the organization. Fire service unions must now ask themselves what values can we offer the fire service.

Rising costs of medical care

As the cost of health care continues to grow each year, many Americans are being fiscal strained. At some point in our lives, virtually every American will need medical care. According to Debra Donahue, "organized labor, representing one in ten Americans, is one of the most influential purchasing groups in the health care industry."[4] The U.S. Bureau of Labor Statistics reports that the U.S. household spent, on average, a total of $2,976 on health insurance, medical services, drugs, and medical supplies in 2008, and the portion of their household budget that consumers spend on health care has gradually risen from about 5.3% to about 5.9% since 1997.[5] The U.S. Bureau of Labor Statistics reported in March 2010 that 93% of union workers had access to medical care benefits compared to 69% of nonunion workers. Of state and local government workers who are unionized, 95% have access to medical care, compared to 81% of the nonunion public sector workers. In the private sector, 92% of union workers have access to medical care, compared to only 67% of nonunion employees. The difference in access health benefits is one reason why union employees are spending more money per hour worked on health care than nonunion employees.[6]

The AFL-CIO reported in 2010 that one in three workers under age 35 has no health insurance.[7] Regardless of whether your organization is unionized, the rising cost of medical services is being felt by both labor and management. The issue of medical expenses may take center stage at the bargaining table—even surpassing wages in importance. Just as in other organizations, labor and management are each seeking creative solutions to maintain or minimize the cost of health benefits to fire departments. The issue of health care costs is expected to remain critical during bargaining for the foreseeable future.

The floodgates opened in 2011 when many states filed legislation to amend—or even to remove—the requirement for states and municipalities to bargain over pensions and health care. The most notable battle was waged in Wisconsin, where steps to minimize public sector collective bargaining rights were taken in order to offset budget shortfalls. Figure 10–2 shows the inside of the Wisconsin State Capitol during a public protect against the proposed changes. As many as 18 other states have filed similar legislation. The impact to labor relations across the nation remains to be seen.

Fig. 10–2. Protestors at the Wisconsin State Capitol, demonstrating against the proposed legislation to change the bargaining laws for state public sector workers

The three general areas that affect health care costs are the Medicare system, health care reform, and health care providers, each of which has some impact on the entire system. At the national level, organized labor has been working with each are to reduce the cost of medical costs. For example, local labor unions may work with the municipal government to implement a wellness program (as discussed in chapter 9) to reduce health care costs.

Another focus has been the implementation of an early retirement age for firefighters. Many reports have shown that older firefighters are at much greater risk of injury than younger firefighters. In the State of Massachusetts, firefighters can retire with a maximum pension at age 55, compared to the many other occupations, where the retirement age has traditionally been 65. With a younger, more in-shape workforce, the utilization rate of the health care system will be reduced; less utilization means lower operational costs and, in turn, less cost to the employer and the employee for medical coverage.

Because many firefighters and employers split the cost of health care, each has a vested interest in controlling costs. Partnerships must be formed between the municipality, the labor union, and the health care provider in order to lower costs. Long-term benefits will be a healthier workforce, health programs that are beneficial to all involved, and the possibility of attracting new talent to the fire department with a solid health insurance program. Whereas wages once served as a deal maker or deal breaker, today the benefit package may outweigh the wages in terms of importance to job candidates and labor unions.

In the public sector, the benefit program offered to employees is invariably much better than the programs offered in the private sector or the employer program that the spouse has access to. In this example, if the municipality offers the best program and high employer contributions, the trend will be for employees to take advantage of this great coverage, as opposed to using their spouses' health insurance. Therefore, the municipality is penalized for having the best possible health insurance for its employees. All things being equal, the risk would be shared equally by public and private sector employers; however, this is not the case. Having the best health care program may attract talented employees, but the municipality also runs the risk of having high utilization of the health care programs.

Health care was has been reserved for discussion in this last chapter because it is likely to continue to be a pivotal issue for the labor-management relationship. In a partnership, each side should work together to make a more efficient system and control costs. The importance of health care costs cannot be overstated.

Strategic planning

The fire service is in a state of constant change—whether small or large. The labor movement is also a living organization, responding to the changing world around it. An organization that is not moving forward grows old, soon to be outdated and doing little for the people it serves. On a local level, by utilizing a strategic plan, the fire department and the local union are thinking strategically, preparing themselves for the future.

The strategic plan is crafted together by the many people who either are part of or are affected by the fire department or the labor union. It is about incorporating peoples' ideas of how they see the fire department or the union into a road map leading to their realization. The strategic plan will allocate physical and human resources to predetermined activities in order to achieve the organizations' goals. As Mark Wallace has explained, the strategic plan is "about increasing an organization's range of future possibilities. It's about building the capacity to respond to anything the organization will face in its future."[8] Strategic planning has been considered a management function, so no matter what type of organization you belong to, the organization could benefit from such a strategic plan.

Labor unions can benefit from having a strategic plan, yet many local fire department unions have not taken the time or the energy to prepare themselves for the future. There are several reasons why the fire service hasn't seen a greater shift toward the adoption of strategic plans. Some fire chiefs have written a department strategic plan one afternoon only to find that what looked great on paper is not practical to implement. A strategic plan takes a long time to develop and a lot of energy but will yield great dividends in the long run.

The strategic plan goes much deeper than identifying the organization's goals and objectives. The strategic plan sets in motion how the organization plans on obtaining these goals and objectives over a period of time. Many departments will establish goals and objectives on an annual basis, only to leave the rank and file and supervisors with no idea as to how to meet those goals and objectives. The strategic plan clears up this confusion

Plan controls—such as budgeting, information systems, and reward systems—are critical to the process of adopting a strategic plan. The incorporation of plan controls will minimize or eliminate factors competing between or against the strategic plan. Hence, the strategic plan is a living document that will need to be adjusted as the organization moves forward to adjust to the changes and obstacles the organization faces.[9]

One common mistake is the failure of an organizations to devote sufficient resources to the overall plan, in part because departments that allocate the resources tend to be short-sighted, expecting immediate results from the strategic plan. Organizations should incorporate the strategic plan in daily decision-making, with the management allowed to make decisions and take action based on the workings of the strategic plan. The rate of return on a strategic plan will be seen over the long run (years), which is frustrating to most of us, who are used to instant gratification in this digital age. Kaplan and Beinhocker have stated that "companies that achieved such success used strategic planning not to generate strategic plans, but as a learning tool to create 'prepared minds' within their management teams."[10] The Phoenix Fire Department has embraced this idea and crafted their strategic plan with cooperation from the union. Thus, the strategic plan represents a combined effort of each side, not just a plan implemented by administration.

Student Activity

The U.S. Fire Administration has a strategic plan for 2010–2014. Review the plan to see how it has mapped out the future of the organization:
http://www.usfa.dhs.gov/downloads/pdf/strategic_plan.pdf

The National Wildfire Coordinating Group has a strategic plan. Review the plan to see how your local labor association could craft a similar plan:
http://www.nwcg.gov/nwcg_admin/strategic-plan.pdf

The New England Division of the International Association of Fire Chiefs also has a strategic plan that you should review:
http://www.nediafc.com/images/pdf/NEDStrategicPlan_2year_2008.pdf

E-Union

By now, today's labor movement should have adopted the *cyber-union* model suggested by Dr. Arthur B. Shostak, Professor of Sociology at Drexel University. Dr. Shostak has observed that, among the bright new generation of emerging union leaders today, "better use is being made of the Internet to weave together electronic social communities of unionists with and across craft and industry line. Union democracy, the finest difference between a labor bureaucracy and a business bureaucracy, is gaining supporters from coast to coast."[11] The cyber-union model takes the past successes of the labor movement and applies them to an online world.

The millennial generation is entrenched in instant information yet has had little material on which to form an educated opinion of the labor movement, so the Web represents a good place to start reaching out to the younger generation (fig. 10–3). The union Web site can have an external section, serving as the public face of the local union, and an internal section for members only. The internal section should allow union members a voice in their union, for instant input regarding members' concerns, without waiting for the next union meeting or speaking to the union steward.

Does your fire department's Web page have a member's-only section? Are the firefighters allowed a voice in the Web page? Can they submit suggestions, recommendations, and complaints via the Web page? Regardless of the nature of your organization (fire department or union), look at your Web page and ask the following: What do the stakeholders and audiences want and need from your Web site, and how will you measure its success? All too often Web designers will put up a Web page and think that it is the best approach to an issue, while failing to implement any means of determining the success of the application.

Fig. 10–3. Instant information via e-union. Almost all unions now have a Web page describing their mission and how to become a member.

Dr. Shostak expanded the cyber-union model in terms of FIT: "futuristics (perspective), infotech (cutting-edge tools), and tradition (commitment). Its appreciation for what FIT can do for it has enthusiasts believing the cyber-union could enable labor to recruit millions of new members and surge forward early in the next century."[12] The message that the labor movement must project is one that will not be jejune to the receiver and therefore must be conveyed in positive manner and in a format that commands attention, as well as be transmitted via various communication means.

The e-union is just one means for making information immediately available to the public and to union members. The union Web page should be selling guests on the benefits of belonging to a labor union. It should not be limited to spreading awareness of the labor unions in the fire service; rather, the Web page should give general ideas in support of the labor movement. As simple as it sounds, many Web sites miss this point. Again, consider who your target audience is.

The International Association of Fire Fighters (IAFF) Web page provides information for the general public, as well as provides password-protected access for union members. Most important, its Web site allows people to join the IAFF. There are many nonunion fire departments that may be interested in joining the union, and in the future, there may be groups of firefighters who have been mistreated and lack the support from management who will need easy access to join a labor union. From a labor standpoint, firefighters joining a union is a good idea; a better idea is firefighters joining the IAFF. When firefighters join other national unions, it splits the focus of a united trade union; however, I will save for another day the argument as to the pros and cons of joining labor unions outside the IAFF or forming independent unions.

The fire department Web page and the union Web page must be maintained with useful and current information. All too often information on the Web is put up but never taken it down, leaving the site cluttered with information and awkward to navigate. The home page should grab the guest's attention without being crowded, as first impressions do make a difference.

The members-only section of the Web page should be password protected to limit access to members of the union, who can register to log on to the site. A registered member should have access to labor news, a section summarizing up-to-the-minute labor events. The information must be published promptly and kept current; for example, in larger unions, a labor news article might be posted daily.

Members may be provided with union e-mail accounts, thereby avoiding the use of the municipal e-mail service, which has the potential of being made public. Several states' attorneys general have issued advisories that the municipal e-mail service is subject to request under the Freedom of Information Act. The process of entering the secure section of the Web page should be easy for members; otherwise, they may be less inclined to return. The site should have a blog for members to post concerns about work or to propose policy changes to the entire membership. This allows members to have a say in the direction of the union, while also allowing union officials instant feedback to dismiss gossip and rumors.

Labor unions should also provide links to other support groups on the Web such as the social, ethnic, sexual, retired, and disability groups. Union meetings can now be broadcasted via streaming video to allow members to log on to a secured server and watch the meeting remotely, allowing firefighters out on injured leave, disabled, or retired to remain active members of the union; previously, these members have felt like outsiders because of their inability to play a direct role in the union. The union Web page should serve both as a means for members to communicate with the union and as a social medium—no different than a physical union meeting.

Suggested information for the union Web site includes the following:

- Collective bargaining agreement, for easy reference to the contract
- Grievance forms, viewable and available for download (in some cases, the written grievance could be filed online via the union Web site)
- Status updates to current grievances and arbitrators' awards
- Legal updates pertaining to labor laws for the state
- Membership rights including the employees' right not to join the labor union

- Conference and training opportunities
- Status of bargaining process and data collection (the union may need a quick survey of the membership on a particular issue for presentation at the bargaining table)

Consideration must be given to the limitations of an e-union approach. Not all people have Web or computer access, in addition to the necessary computer and Internet skills. Consider a firefighter who is 60 and one who is 20: Which will be more likely to master the e-union Web site and all the possibilities that it presents? This is a great opportunity for the union to partner with a local computer company to offer computer classes and, possibly, deals on computers.

Some argue that the cost of and time spent on developing an e-union puts it out of reach; however, printing, distributing, and other costs associated with physical documentation will be spared, bringing the expense to a more manageable level. As discussed in chapter 6, arbitration has also shifted to an online approach in an effort to streamline the process and to lower the cost. In addition, online education and training are on the increase across the nation.

Social media have been kicking around for the past few years, and many people have now joined some type of social network. Organized labor can use social media to keep members and the public informed about issues. The fire department can also embrace social media. However, this means that they must trust employees to act responsible on social media Web sites, as they are a reflection of the department and the badge. Some departments have attempted to place significant restrictions and policies on use of social networking applications at work or on department computers, forcing people to seek alternative methods (e.g., cell phones and personal computers). As Fiona Powell has reported, "by the year 2013, 43 percent of global mobile internet users (607.5 million people worldwide) will be accessing social networks from their mobile devices."[13] Department policies preventing firefighters from using social networks practically necessitates a bureaucracy of its own. While a fire department cannot stop the use of social networks, some attempt to limit use. However, progressive fire service leaders will inevitably embrace the use of social media (wikis, blogs, forums, and instant messaging) within the agency. Fire service training divisions should be looking at how social networking sites can help them achieve the goal of keeping members up to date on the most current training issues.

Rate of Return on Investment in Labor Relations

A person who is armed with knowledge and understanding of the past (e.g., see fig. 10–4) is better able to forecast the future. Regrettably, many fire service administrators hold a given position only for a short period, before moving on to another department or retiring. Consequently, many fire service administrators do not invest time and energy into building a stronger labor relationship with the labor union.

Fig. 10–4. Rosie the Riveter, promoting women joining the workforce to support our troops during the Second World War

In a fire department that has poor labor relations, new fire administrators often create and implement short-term improvements in the labor relations process during their first three years, referred to as the honeymoon period. After the honeymoon period, the administrator may strive to hold the relationship together or may develop an us-against-them approach.

To build a stronger relationship, labor and management will need to reflect on their own actions and attitudes. While this may sound easy, it is by far the greatest challenge facing each side; it is way too easy to point the finger and blame the other side. Self-reflection may need the assistance of a third-party coach, who will help the organization identify those actions that have had a negative impact on the labor-management relationship. The truth may hurt but is necessary so we can grow. Through self-reflection, labor and management will be better able to take corrective actions to advance the relationship.

The labor-management relationship is, like the fire equipment we use on a daily basis, in need of regular maintenance. When the old equipment must be replaced, we look for the newest and most effective means possible to accomplish the task. Consider the energy and resources we place on building the next fire engine or ladder truck; yet, most fire service organizations will not apply the same level of energy and resource to building a stronger relationship. Equipment is prone to break down; we can choose to fix the equipment, or we can drive it into the ground and destroy the equipment. The same can be said for some labor-management relationships: the two sides may refuse to perform any preventative maintenance or repair the damaged relations and instead choose to drive the relationship into the ground.

In spite of all the talk and research on how organizations must increase productivity and do more with less, many fire departments overlook the waste of expending resources and efforts on a negative labor relationship. Self-reflect on your own organization: How much time, resources, and energy is spent on the labor relationship? Do any area fire departments have a negative relationship? How much is lost by maintaining this negative relationship? Could the time, energy, and resource be better used in a more productive manner?

The desire to continue building the relationship or the desire to take long-term action to build on the relationship is minimal at best. Taking a long-term approach to building/maintaining labor relations will often take the following path:

- *Developing.* This is when labor and management work together to develop ideas that will build the relationship and the labor relations process. This is more about mutual gains.
- *Structure.* The structure should be evaluated to ensure that the labor relations process remains productive today. The structure that worked 10 or 25 years ago may not work today. Labor and management must self-reflect on the system that they each use internally to see if adjustments are needed to meet today's challenges. Adjustments may be needed to the structure to strengthen the relationship; each side must look outside the box and be willing to take the risk of doing something different.
- *Process.* The process that is used to administer the bargaining process and the administration on the collective bargaining agreement should be evaluated and adjusted to best meet the current needs. Labor and management should self-reflect on the internal process they use in the labor relations process to see how they can adjust.

Today many fire departments have learned a harsh lesson on preparing for the long term. With the economic crisis, many are still in the recovery process. As the economy changes, labor unions and the fire service collectively will need to change the way we have done things in the past, to face new demands. Sitting at the kitchen table in any firehouse, firefighters can offer an opinion as to where the fire service is heading; some may be right, while others may be wrong. If we are not looking forward and assessing threats, we will find ourselves in a reactive state, allowing the threat to manage us. If you are an aspiring fire service administrator, you should keep an eye on the future so that you and your organization are better prepared to adjust—ideally, before the change demands that an adjustment is needed. Otherwise, the change will manage us, as opposed to our managing the change. When a fire department or a labor union must make adjustments based on what they are currently facing, they are playing catch-up, which will often require that a greater level of change be imposed over a much shorter time period. This reactive state will produce long-term stress on personnel and the organization. In particular, personnel tend to be more resistive to changes made on short notice and as a result of current demands.

As time passes, the public sector labor relations process may evolve to become more standardized, as states draft legislation to bring their laws in line with each other. Even streamlining a process that is currently operational may be insufficient, as today's fire service serves different needs than it did 25–30 years ago when the process was implemented. This communality in states will be minor. We are still years away from a national standard for public sector bargaining laws, as the states are not about to release their power over public sector bargaining. Some level of standardization may be helpful and would even the playing field for all public sector employees; however, this is not the magic bullet to fix the labor relations process in the public sector.

Organizational ethics

A firefighter faces ethical choices every day, whether on the job or off duty. Is it wrong for the firefighter to be writing a college paper on duty and on the department computer? Is it wrong to print the report on the department printer with the department's paper? Is it wrong to withhold data for a bargaining session? Is it wrong to be conducting union business on duty? As you can see, a number of factors come into play in answering these questions. Is it wrong for a instructor to let students out early for a mandatory training session required for certification as an emergency medical technician (EMT)? And is it ethical? Is the instructor, knowingly or unknowingly, breaking the law or violating a legal mandate?

We must first define ethics. As Gary Dessler has stated, "ethics refers to 'the principles of conduct governing an individual or a group.'"[14] The conduct governing the individual or group may be in the form of a guideline, rules, or standards, or it could be established by the group. When we talk about personnel issues, one may need to consider the individual's values, cultures, religious beliefs, and personal biases. Different people bring different things to the table, and we should be respectful of these differences, even while that does not provide them with a blank check to do what they want. The group—the fire department or even the union—could establish the acceptable standard. There may also be subgroups, composed of each of the shifts within the fire department or specialized units. From a religion standpoint, ethics may be based on the 10 Commandments, and the legal system creates laws reflecting the changing social values and attitudes of our country to determine whether wrong has been committed.

The ebb and flow of social customs can also determine ethical behavior and to whether society will condone or condemn any given action.[15] Ethical behavior is fluid and can shift over time—what was considered acceptable 25 years ago may no longer be. Hence, firefighters must have a solid understanding of ethics, and ethics must be at the heart of any discussion on the future of the fire service. Regardless whether we are in the back of an ambulance, in the firehouse, in the midst of labor negotiations, or administering personnel, we all face ethical challenges.

Regrettably, in the past, fire departments often neglected the issue of ethics until an ethics violation was investigated, and labor unions did not make an effort to establish a code of ethics. Whenever you turn on the television or read the newspaper, there seems to always be some business leader or worker accused of an ethics violation. Ethics violations are ubiquitous in today's business world, and the fire service is no exception. Bernie Madoff, the poster child for ethical violations in business, created a Ponzi scheme, siphoning billions of dollars from clients who trusted and respected him as a professional. On his own, Mr. Madoff made the decision to engage in this unethical behavior.

Experts in the field of ethics have recommended that fire departments create an ethical standard as to what behaviors and conducts will be condoned or condemned. Further, the labor organization should also create an ethical standard. Some departments have taken this advice, only to find that they are still troubled by unethical behavior. This is because the organization needs to ensure that each level of the organization—from firefighter to fire chief—has embraced the boundaries established by the organization. Training and awareness must be incorporated into the process. In addition, the ethical standards need to be reviewed more than on an annual basis; rather, this should be incorporated into the daily practice of the department. The department that incorporates the above recommendations stands the greatest chance of the having ethical boundaries become accepted as the norm of the department.

Firefighters should not have to ponder what is right or wrong; as long as the organization has adopted and embraced an ethical standard, the right course should come naturally to the firefighter. When determining an ethical standard, you will always have people on each side of the issue, and the challenge is to find common ground as to what is and is not acceptable. Discussion of ethics is often controversial, as there are many processes a person or group may use to arrive at an ethical decision—no one theory is the best.

An argument over ethical decisions sometimes results because the individual/party is so entrenched in defending a predetermined position and demonizing the other individual/party (posturing) that they lose perspective to on ideas outside their position. Therefore, it is essential to try and understand the other person's position, as well as to understand their resistance to your position, in order to reach some mutually acceptable position. Do not argue against the person or the position; instead, attempt to think as the other individual/party. David Arnaud has elaborated on this point:

> To dialogue with others, by genuinely acknowledging the possibility that others might have wise things to contribute to our thinking, even when they disagree with our present views, requires a radical change in our current practice. Even this, radical as it is, is at most a necessary, rather than a sufficient, condition for wiser ethical decision-making, as a willingness to dialogue without a method for dialoguing well is likely to be severely limited in its effectiveness.[16]

When we involve people in the decision-making process, there will more often than not be differing opinions. This contributes to a healthier organization, interested in fleshing out ethical issues. The establishment of ethical standards should not be a stand-alone process, rather it is an integral part of a bigger process. In the diagram that Leanee Hoagland-Smith describes (fig. 10–5), the core values would encircle the actions of the fire department, and inside the circle are the fire department's vision, mission, critical success factors, goals, action steps, all of this rests on the one word: *results*. In the simplest terms, only decisions that have been predetermined as acceptable positive behaviors should be selected. This model can be adapted to the individual fire department or labor union.[17]

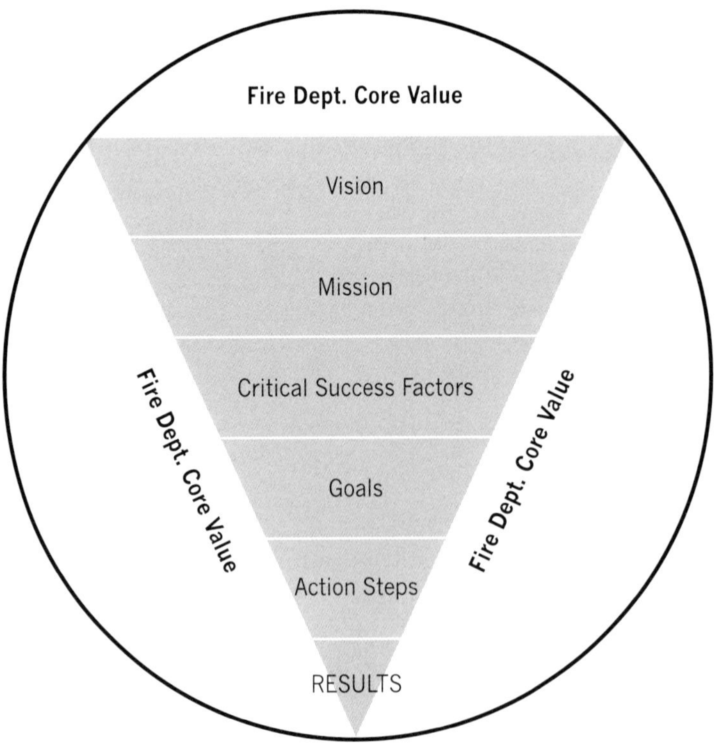

Fig. 10–5. Ethical decision-making diagram

As fire chief, one of my first priorities for my department was to determine how we would move forward as an organization in a progressive manner. Members of the department wanted a road map of how we would move forward. In response, I would often say, "We will REACH into the future." This is an easy-to-remember acronym for the following ethical model:

- *Responsibility.* We all have a responsibility to the department, ourselves, and our family.
- *Ethics.* We will conduct ourselves in an ethical manner.
- *Attitude.* We will present a positive attitude.
- *Choice.* When faced with choices, we will identify and support the correct choices.
- *Health.* We will take action to promote the health and safety of the community and the department.

I had 150 coffee mugs made with the impression of the REACH motto (fig. 10–6). When I sat down with individual firefighters, town officials and department heads, and area fire chiefs, I gave them each a mug and explained how the department would move forward. As the coffee mug was used by department members, who thus saw the REACH motto on a daily basis, I could ask the members how issues the department faced fell within the REACH model. Regardless of whether it was a personnel issue, an operational issue, or an equipment issue, the employee could come up with a decision based on the REACH model. From this starting point, we moved forward in redesigning the department polices and procedures in cooperation with the labor union.

One member wanted to design a customized EMT patch for the uniform that would be respectful to the community. The town seal was set in the center of the patch with the words "Certified Emergency Medical Technician" around the outer edge. The patch was incorporated into a mandatory insignia for the uniform, replacing the standard state EMT patch. With the patch located just above the right-side chest pocket, a patient being treated could easily see that the person administering the treatment had obtained the level of training represented by the title of EMT.

Whenever you incorporate members in the decision-making process, resistance will be minimized. When members can see that solid ethical decisions are being made and supported by the administration, they will tend to behave accordingly. Praise acceptable behavior so that others can see what is correct. Correct the unacceptable behavior in a teaching fashion if possible, before taking any disciplinary action. When preparing my annual town reports, I would often refer to the REACH model to illustrate the milestones that the department had already achieved and had not yet reached.

Fig. 10–6. The REACH coffee cup, given to each member of the department and to all the elected officials of the town. The coffee cup served as a daily reminder of how we would reach into the future.

Fire service culture

The fire department culture and the union culture may be in line with one another, or there many be cultural differences, resulting in turmoil. Culture, more or less, can be defined by the behaviors and knowledge that is considered by the group of individuals as social behavior. A fire department or union's culture is learned, so that individuals may communicate with each other in the group and externally. The department culture will play a central role in the way that the personnel management system will be used. Newly hired firefighters may have a difficult time adjusting to a fire department or union that has different values and beliefs. There are three choices the new firefighters can make:

- Leave the department
- Learn and accept the department and union values and beliefs
- Retain their own values and beliefs without conforming to the department culture

The path of least resistance is the first, to seek employment elsewhere, with an organization that holds values and beliefs more in line with those of the individual firefighter. The second choice, in which the new firefighter suspends or alters (for good or bad) his/her values and beliefs to be more in line with the organization, is known as *collectivism* or *group orientation*. The energy-filled new hire quickly realizes that others in the organization behave differently and learns to act as the others do. In terms of personnel management or labor relations, this new firefighter, who was initially eager to learn and take on new challenges, may soon become another individual in the group by taking on the cultural norms; hence, in a department or union with a substandard value/belief system, a new hire may take on the substandard values/belief system. Depending on how entrenched the culture is within the organization, the administration's attempts at change can be significantly hampered. Because the fire service is so dependent on group-oriented interaction, firefighters are generally evaluated by the group as trustworthy, loyal, and compatible with the group.

The third choice, to stand firm on one's own values/belief despite differing from the organization's, is known as *individualism*. In this case, the new hire may not be socially accepted into the organization as a result of the cultural differences and thus may be considered the misfit of the organization. The person may be subjected to harassment and intimidation to conform to the organization's culture. Some firefighters who select this option may be able to withstand the social pressure to conform to the organizational culture, whereas others will eventually break down and either leave or become disgruntled and unproductive. As an aspiring fire service administrator, one should be aware of this force with the organization. Left unchecked, the organizational culture can have a negative impact on the organization.

Importantly, the organizational culture is learned, so a forward-thinking fire service administrator can and should influence the organization to unlearn negative values/beliefs. Although it can be done, it will take time. To be successful, the fire service administer should have an understanding of the values/beliefs held by the fire department and the labor union. The fire service administrator should be sensitive to deeply held beliefs that are not counterproductive, negative, or illegal.

Ensure that the ideas (ethical behavior) are meaningful and lasting, not just a flash in the pan, forgotten in a couple of weeks. If you are a fire service leader, then your behavior will set the proper image (*inspiration*) for the employees that you are supervising; the teaching and modeling (*facilitation*) process starts with the chief of the department. The chief will establish the proper behavior; once employees see what is expected, the acceptable behavior will trickle down to the rank and file by way of supervisors and managers. As Randy Bruegman has stated, "the leader must employ the authority granted him or her by the organization to achieve the purpose of the organization, while recognizing the knowledge needed to exercise this authority resides throughout the organization and its environment."[18] Reversing the process is more challenging, as issuing a policy or directive against a specific behavior will very likely be met with resistance. It is far easier for the chief of the department to establish the proper behavior and take small incremental steps in the right direction than to take one major step against the tide.

Over the past few years, the fire service has transitioned into specialized units such as the Hazmat team, high-angle rescue, SWAT medic program, dive rescue team, and fire prevention. Member selection is conducted a number of ways, from appointment by the chief to nomination by the current team members. Highly specialized units develop a culture of their own, which is a subculture of the department culture, and the specialized unit generally will develop a sense of superiority, referred to as *group centrism*. Collectively, the fire department also projects group centrism—

after all, we are the best at everything we do and better than the other areas fire departments. First and foremost, group centrism serves to promote and protect our own specialized unit. If the specialized group is left unchecked, to run autonomously, problems can be encountered down the road as this specialized group continues to develop a cultural standard (good or bad) among the team. Patrick O'Hara has explained, "cultural deviation occurs when a group of employees operate in disregard of organizational rules and official norms as matter of daily process and moral value. When this condition is acute, cultural deviant elements of an organization operate largely unto themselves and for themselves."[19]

The assimilation of newly hired firefighters is not the only cultural challenge the organization will face. When a fire department chooses to go outside the department to hire a deputy chief, assistant chief, or chief of department, the community must take into consideration the culture from which the individual is coming and determine whether he or she will fit into the department culture. The fire chief, when hiring a subordinate, should take steps to ensure that the person will fit not only the culture but also the management style that is in place. If a person has worked for only one fire department during his/her 25-year career, the issue of adjustment may be a concern that the hiring team should weigh carefully. Candidates who apply for a job outside the department after 15–25 years run the risk of cultural and administrative adjustment issues.

In the human resource field, the old idea was that a person who jumps around may not be a good hire; however, when a fire department is looking to hire from outside, it may want to look at candidates who have had the benefit of working for and advancing through different organizations in the fire service. Another consideration is the ability of a union firefighter to be promoted outside the union and serve on the management team. Can the candidate adjust to working for management? Will he or she become too involved with the rank and file? Don't get me wrong—some very good fire chiefs have come directly out of the union ranks. In fact, this can be an added benefit as they know how to work in a union environment. Yet the question remains whether they can make the transition to management smoothly. The candidate who has been promoted to management within the same department may find the transition to be more challenging, as compared to a candidate from outside the department who is transitioning into management.

One community made a costly mistake hiring the wrong person for the position of assistant chief. After a couple of months on the job, it was apparent that the new assistant chief was not the right fit for administration of the department and held too close a relationship with the local union. The latter was the more concerning, because a jaundiced relationship between labor and management resulted. Even with all the effort that the town had expended—four months, countless hours, and more than three interviews conducting the search, as well as input from outside the fire department—the wrong person was hired. The good news for this community was that the newly hired assistant chief was able to leave and return to his prior job with only a very small severance package. The moral of the story is that the organizational culture that a firefighters comes from may make or break a person. The hiring decision should be based not on any single issue, but instead on many small factors.

Training and education

The discussion of training and education has been placed in this chapter because it should also continue to take center stage in the future. Training can range from firefighter orientation to continuing education on skills. No matter how much time a fire department spends on the selection of firefighters, there is no guarantee that the person can perform effectively (fig. 10–7).

Fig. 10–7. Firefighter training and education. All ranks/grades within the fire department require some level of training and education to ensure peak performance. Each firefighter within the department should know what to do and how to do it.

In the old days, a company officer provided a brief training session in the classroom or in the field. The problem with this type of training was that it rested on the company officer to come up with the training, and consequently, there was no consistency, even between shifts. In addition, there was the potential for legal trouble because there was no assurance that fire service standards were being met by this type of training. However, today, many commercial training programs are available for purchase, some of which provide lesson plans on a given topic, teaching tips, and quizzes to assess the level of learning achieved. A commercial fire service training program will often list the objectives to be met and the standards used, in case training is called into question in a civil lawsuit.

The training and development process follows five steps. The first step in the process involves conducting a *needs analysis*, to identify the specific job skills needed, the firefighter skill level, and areas of need for training on the basis of past performance and injuries. During the analysis stage, the measurable knowledge and performance objectives should be based on deficiencies. The second step is the *instructional design process*. During this process, data and research are compiled and the decision is made on how the material will be presented (in the classroom, in the field, by independent study, or online). The third step is the *validation process*, often called a *pilot program* when the proposed training program is presented to a small group of people to determine whether the material addresses the intended audience. The program and activities are checked against time requirements, to ensure that they run smoothly. At this point, adjustments may be made to the program. The fourth step is *implementation*, during which the program is rolled out to the intended audience (e.g., the entire department). The fifth step in the process is the *evaluation* of the impact that the training program had on the intended audience. The evaluation methods should incorporate measurable performance indices.

The most common problem with fire department training is that people tend to omit the first and/or the fifth step, with disastrous results. If you skip the first step, then you have little to no knowledge on which to build your training program. If you skip the final step, then you will never know whether your training program accomplished what was intended.

The fire service is finally starting to embrace ideas outside its prior comfort zone of classroom presentation and field skills development, including distance and Internet-based training. In larger fire departments, the training officer may be at the fire academy, and tele-training is presented in several fire stations distributed across the city. Another option is to use videoconferencing to conduct meetings or hold discussions. This will allow companies to stay in service with their districts, rather than all coming to headquarters. Almost all fire stations now have computer terminals and Internet access, so implementing a training or discussion session would be easy and cost-effective, so that the training officer could present material only once with all members receiving the same message, as opposed to going to each station and conducting the same training session several times. Sessions can even be videotaped to allow members to review a section missed because of an emergency call, for example. Likewise, several commercial Internet-based training programs are available to the fire service.

A department may partner with a local college to deliver a management class to all fire officers; this could be an online class in which each student logs on to the class and completes the course within a set time frame. A county or regional area may find this type of training cost-effective because the material can now be used in several departments, as each department will not have to pay the training officer separately for the exact same class. Many community colleges now offer EMT refresher classes online, with one day set aside for a physical practice or drill session. Again, this is a cost-effective means of delivering training material to a large audience. In addition, it may be convenient for the individual firefighter. As Gary Dessler has stated,

> Researchers reviewed 96 studies reporting data from 19,331 trainees who participated in 168 training courses, both Web-based and classroom. In general: Web-based instruction was a bit more effective than classroom instructions for teaching declarative knowledge (memory of facts and principles; Web-based instruction and classroom instructions were equally effective for teaching procedural knowledge (information about

> how to perform a task or action); trainees were equally satisfied with Web-based instruction and classroom instruction; and Web-based instruction was much more effective than classroom instruction when the trainees could use the Web programs to control the pace and selection of the content.[20]

The best approach to training and education is to include a blend of Web-based training and classroom learning, to hit the different learning styles of students. There should be no fear that Web-based training will supplement traditional classroom instruction. Labor and management should create a partnership to create an enhanced training and education program for the department. Involving the people who will be receiving the training in the development stage of the training program will yield the greatest dividends.

Business partner

Labor unions in the public sector need to become a business partner of municipal government. Much more can be obtained when the two sides work in a partnership toward a common goal. This means rethinking how we have done things in the past; together, we need to better define how government can advance productivity, increase quality, and improve government competitiveness (privatization is a real and big threat to some levels of government). This means that labor may take on a large role in the decision-making process from a business standpoint. The union and management would equally take on the risk and the rewards of a successful organization. Each would be concerned with the bottom line of the budget, and each would retain a responsibility to maintain a successful business. Unfortunately, few in the public sector use a true business approach. Streamlining government is a must today; thus, labor and management should partner to reengineer government. State or national labor unions representing firefighters could be underwriting training programs for young people interesting in becoming firefighters, and organized labor could be partnering with local colleges or technical colleges to help underwrite such training programs. Labor and management could form partnerships in creating training and education for fire service managers and aspiring fire service administrators. This would be much more than a union drive for new members; rather, it would set the stage for the best talent to be in the fire service, with a vested interested in the risk and benefit of a business partnership. Creating efficiency and competitiveness through a business partnership should be the goal.

Final Thoughts

The preceding chapters may have either been new material to you or served as a refresher. Regardless of your knowledge of labor relations, you have been shown the positive and negative sides of the labor relations process, and the examples used in this book may stick with you as you continue to learn. The student activities in each of the chapters have been designed to expand your knowledge on the subject matter and encourage you to examine your own individual department, to enhance the labor relation process, as well as the department personnel management system. If you are an aspiring union official or fire service administrator, you can see the need for understanding the labor relation process and how it is interconnected to personnel management.

The goal of this book has been to provide you with an understanding of the labor relations process for the fire service. As a current labor member or administrator, this may help you to identify concerns within your labor relationship and personnel management system. To build a stronger fire service, labor and management must continue to cooperate in every aspect of organizational development. Organizational performance plus improved decision-making results equal organizational development.

I have suggested avenues to correct problems in personnel management and to overcome long-standing issues between labor and management. It is your responsibility to put these suggestions to work in your department. No matter how big or small your department is, you should bring positive change to the fire service. Do not just put this book back on the bookshelf; instead, use it to bring about change in labor relations and personnel management.

Many of the principles and practices suggested in this book have direct roots to the human resource field and personnel management. Labor relations and human resources management have strikingly similar tasks. If the suggestions in this book can help one fire service agency, labor union, or employee, then all of the hours spent writing and compiling data and research will have paid off. It is my hope that this book will also serve as an education component of senior management training programs, as well as courses in the higher education piece for the fire service.

There is no right or wrong way for organized labor to move forward. Rather, the process involves a number of pieces that make up the greater puzzle of labor relations; remove one or a couple of pieces of the puzzles, and it still would work, although the true picture would not be as clear. To make this work, consider the foundation of the labor relations process: efficiency, equity, and voice. These three items have been emphasized throughout this book. If fire service leaders and union leaders can balance these items, then the process of labor relations will progress in a positive manner. Positive labor relations means a better work environment; a better work environment in turn means happy employees; and finally, happy employees open up unlimited possibilities.

Key Terms

- Collectivism
- Cyber-union
- E-union
- Ethical decision making
- Fire service culture
- Group orientation
- Group centrism
- Individualism
- Instructional design process
- Millennial generation
- Needs analysis
- Organizational ethics
- Pilot program
- Social media
- Strategic planning
- Validation process

References

1. Budd, J. 2008. *Labor Relations: Striking a Balance*. 2nd ed. Boston: McGraw-Hill Irwin. P. 498.
2. Parks, J. 2009. Diversity summit: Future of unions depends on including all workers. AFL-CIO Now Blog. http://blog.aflcio.org/2009/09/13/diversity-summit-future-of-unions-depends-on-including-all-workers/ (accessed September 10, 2010).
3. U.S. Department of Labor, Bureau of Labor Statistics. 2010. Union members—2009. http://www.bls.gov/news.release/union2.nr0.htm (accessed September 10, 2010)
4. Donahue, D. 2010. Health coverage for labor union markets. http://www.markfarrah.com/healthcarebs.asp?article=75 (accessed September 5, 2010).
5. U.S. Department of Labor, Bureau of Labor Statistics. 2009. Health care. http://www.bls.gov/spotlight/2009/health_care/ (accessed September 5, 2010).
6. U.S. Department of Labor, Bureau of Labor Statistics. 2010. Employee benefits in the United States, March 2010. http://www.bls.gov/news.release/pdf/ebs2.pdf (accessed November 25, 2011).
7. AFL-CIO. 2010. Young workers—a lost decade: One year later. http://www.aflcio.org/aboutus/youthsummit/upload/young_workers_100810.pdf (accessed October 26, 2010).
8. Wallace, M. 2006. Fire department strategic planning: Creating future excellence. 2nd ed. Tulsa, Okla.: PennWell.
9. Gary, D. 1986. Use and misuses of strategic planning. *Harvard Business Review*. January/February, 64(1), p. 89-97.
10. Kaplan, S., and E. Beinhocker. 2003. The real value of strategic planning. *MIT Sloan Management Review*. 44 (2): 71–76.
11. Shostak, A. 2010. Personal communication. October 18.
12. Shostak, A. 1997. Organizing via CyberUnion: Can 21st century cyberunions be created in time? http://www.futureshaping.com/shostak/pages/essayCyberUnion.html (September 1, 2010).
13. Powell, F. 2010. Social media and employees. *Human Resources*. April/May: 24–25.

14. Dessler, G. 2008. *Human Resource Management*. 11th ed. Upper Saddle River, N.J.: Pearson-Prentice Hall.

15. Bruegman, R. 2009. *Fire Administration I*. Upper Saddle River, N.J.: Pearson-Prentice Hall. P. 236.

16. Arnaud, D. n.d. Using a decision-diagram for ethical decision-making. http://www.decision-making.co.uk/Publications/DecisionDiagram.pdf (accessed September 19, 2010).

17. Hoagland-Smith, L. n.d. Success in business ethics is about internal behaviors not external laws or regulations. http://ezinearticles.com/?Success-in-Business-Ethics-is-About-Internal-Behaviors-Not-External-Laws-Or-Regulations&id=3627530 (accessed September 10, 2010).

18. Bruegman 2009, p. 247.

19. O'Hara, P. 2005. Why law enforcement organizations fail: Mapping the organizational fault lines in policing. Durham, N.C.: Carolina Academic Press.

20. Dessler 2008, p. 308.

INDEX

Case Index

Subject Index

A

B

D

E

F

G

H

M

R

S

T

U

V

W

Z

ABOUT THE AUTHOR

Paul Antonellis, MA, has more 20 years of fire/EMS experience, holding various positions including Chief of Department. He served as the Vice President of the Yarmouth (MA) Professional Firefighters Union and was a founding member of the Professional Fire Fighters of Massachusetts (PFFM): Members Assistance Committee (MAP). He traveled within the state of Massachusetts providing assistance to brother/sister union members as a Certified Employee Assistance Professional (CEAP), Certified Addictions Specialist (CAS), and a licensed alcohol counselor. Antonellis served as the Coordinator of Municipal Support at the Massachusetts Firefighting Academy and a National Examiner for Firefighter I/II and Instructor I. In addition, Antonellis has 11 years of law enforcement experience with the Dennis (MA) Police Department as a special police officer.

Antonellis is on the faculty of three colleges teaching in the undergraduate and graduate programs, in addition to working in the curriculum development division. He has taught more than 70 undergraduate and graduate level courses over the past seven years. He served as a subject matter expert for the Labor Relations class and the Emergency Management graduate classes at Southern New Hampshire University. He teaches extensively in the Organizational Leadership Program and the Justice Program. He also teaches at Empire State College (Saratoga Spring, NY) in the Fire Science Program and Labor Studies Program. Antonellis served as a consultant in the development of the online fire science program at the college and is a subject matter expert for several classes. He has the experience of teaching in the classroom and online. Paul was also tapped by the National Fire Academy to serve as the team leader for the development of the Degrees at a Distance class Fire Research. He was responsible for the management of the curriculum development team in crafting this new online format of learning for the fire service. Several colleges around the country have accepted the Fire Research class at the undergraduate level in partnership with the National Fire Academy and meeting the FESHE requirements.

He has lectured to emergency service providers nationally and internationally on various topics. Paul has appeared in several national training videos, including a line-of-duty-death training video. Antonellis has served as a training/education consultant to the U.S. Coast Guard, U.S. Air Force Reserve, U.S. Probation Department, East Hartford (CT) Police Department, District of Columbia Fire and Emergency Medical Services, North Dakota APCO Association, Atlantic Chapter—Association of Public Safety Communications Officials, and many municipal fire departments in the state of Massachusetts. He has authored and published more than 25 articles on various emergency service topics and has been published in *Fire Chief Magazine*, *Firehouse*, *Fire Engineering*, *Public Safety Communications*, *FireRescue Magazine*, *Law Enforcement Employment Bulletin*, *Law Enforcement Bulletin*, *Police Department Disciplinary Bulletin*, *The Voice*, *Firestation Management Advisor Newsletter*, and for the Association of Public-Safety Communications Officials. Antonellis authored and published two books, *A Guide to Help the Fire Service Prepare for a Line of Duty Death or Injury* and coauthored the book *Posttraumatic Stress Disorder in Firefighters: The Calls That Stick With You*.

His education includes a masters degree in labor and policy studies with a concentration in human resource management, a bachelor of science degree in fire service administration, an associate of science degree in criminal justice, and an associate of science degree in fire science technology. As a member of the ICISF faculty, he teaches several programs including "Line-of-Duty Death: Preparing the Best for the Worst," which he codeveloped.